NANOTECHNOLOGY SCIENCE AND TECHNOLOGY

TiO₂ NANOCRYSTALS

SYNTHESIS AND ENHANCED FUNCTIONALITY

NANOTECHNOLOGY SCIENCE AND TECHNOLOGY

Safe Nanotechnology
Arthur J. Cornwelle
2009. ISBN: 978-1-60692-662-8

National Nanotechnology Initiative: Assessment and Recommendations
Jerrod W. Kleike (Editor)
2009. ISBN: 978-1-60692-727-4

Nanotechnology Research Collection - 2009/2010
James N. Ling (Editor)
2009. ISBN: 978-1-60741-293-9
(DVD edition)
2009. ISBN: 978-1-60741-292-2
(PDF edition)

Strategic Plan for NIOSH Nanotechnology Research and Guidance
Martin W. Lang
2009. ISBN: 978-1-60692-678-9

Safe Nanotechnology in the Workplace
Nathan I. Bialor (Editor)
2009. ISBN: 978-1-60692-679-6

Nanotechnology in the USA: Developments, Policies and Issues
Carl H. Jennings (Editor)
2009. ISBN: 978-1-60692-800-4

Nanotechnology: Environmental Health and Safety Aspects
Phillip S. Terrazas (Editor)
2009. ISBN: 978-1-60692-808-0

New Nanotechnology Developments
Armando Barrañón (Editor)
2009. ISBN: 978-1-60741-028-7

Electrospun Nanofibers and Nanotubes Research Advances
A. K. Haghi (Editor)
2009. ISBN: 978-1-60741-220-5
2009. ISBN: 978-1-60876-762-5
(E-book)

Carbon Nanotubes: A New Alternative for Electrochemical Sensors
Gustavo A. Rivas, María D. Rubianes, María L. Pedano, Nancy F. Ferreyra, Guillermina Luque and Silvia A. Miscoria
2009. ISBN: 978-1-60741-314-1

Polymer Nanocomposites: Advances in Filler Surface Modification Techniques
Vikas Mittal (Editor)
2009. ISBN: 978-1-60876-125-8

Nanostructured Materials for Electrochemical Biosensors
Yogeswaran Umasankar, S. Ashok Kuma and Shen-Ming Chen (Editors)
2009. ISBN: 978-1-60741-706-4

Magnetic Properties and Applications of Ferromagnetic Microwires with Amorphous and Nanocrystalline Structure
Arcady Zhukov and Valentina Zhukova
2009. ISBN: 978-1-60741-770-5

Electrospun Nanofibers Research: Recent Developments
A.K. Haghi (Editor)
2009. ISBN: 978-1-60741-834-4

Nanofibers: Fabrication, Performance, and Applications
W. N. Chang (Editor)
2009. ISBN: 978-1-60741-947-1
2009. ISBN: 978-1-61668-288-0
(E-book)

Bio-Inspired Nanomaterials and Nanotechnology
Yong Zhou (Editor)
2009. ISBN: 978-1-60876-105-0

Nanotechnology: Nanofabrication, Patterning and Self Assembly
Charles J. Dixon and Ollin W. Curtines (Editors)
2010. ISBN: 978-1-60692-162-3

Gold Nanoparticles: Properties, Characterization and Fabrication
P. E. Chow (Editor)
2010. ISBN: 978-1-61668-009-1
2010. ISBN: 978-1-61668-391-7
(E-book)

Micro Electro Mechanical Systems (MEMS): Technology, Fabrication Processes and Applications
Britt Ekwall and Mikkel Cronquist (Editors)
2010. ISBN: 978-1-60876-474-7

Nanomaterials: Properties, Preparation and Processes
Vinicius Cabral and Renan Silva (Editors)
2010. ISBN: 978-1-60876-627-7

Nanopowders and Nanocoatings:
Production, Properties and
Applications
V. F. Cotler (Editor)
2010. ISBN: 978-1-60741-940-2

Barrier Properties of Polymer Clay
Nanocomposites
Vikas Mittal (Editor)
2010. ISBN: 978-1-60876-021-3

Nanomaterials Yearbook - 2009.
From Nanostructures,
Nanomaterials and
Nanotechnologies to Nanoindustry
*Gennady E. Zaikov and
Vladimir I. Kodolov (Editors)*
2010. ISBN: 978-1-60876-451-8

Nanoparticles: Properties,
Classification, Characterization,
and Fabrication
*Aiden E. Kestell and
Gabriel T. DeLorey (Editors)*
2010. ISBN: 978-1-61668-344-3

Nanoporous Materials: Types,
Properties and Uses
Samuel B. Jenkins (Editor)
2010. ISBN: 978-1-61668-182-1
2010. ISBN: 978-1-61668-650-5
(E-book)

Silver Nanoparticles: Properties,
Characterization and Applications
Audrey E. Welles (Editor)
2010. ISBN: 978-1-61668-690-1
2010. ISBN: 978-1- -61728-062-7
(E-book)

Mechanical and Dynamical
Principles of Protein Nanomotors:
The Key to Nano-Engineering
Applications
A. R. Khataee and H. R. Khataee
2010. ISBN: 978-1-60876-734-2

TiO2 Nanocrystals: Synthesis and
Enhanced Functionality
*Ji-Guang Li , Xiaodong Li
and Xudong Sun*
2010. ISBN: 978-1-60876-838-7

Nanomaterial Research Strategy
Earl B. Purcell (Editor)
2010. ISBN: 978-1-60876-845-5

Magnetic Pulsed Compaction
of Nanosized Powders
*G.Sh Boltachev, K.A Nagayev,
S.N. Paranin, A.V. Spirin
and N.B. Volkov*
2010. ISBN: 978-1-60876-856-1

Nanostructured Conducting Polymers and their Nanocomposites: Classification, Properties, Fabrication and Applications
Ufana Riaz and S.M. Ashraf
2010. ISBN: 978-1-60876-943-8

Phage Display as a Tool for Synthetic Biology
Santina Carnazza and Salvatore Guglielmino
2010. ISBN: 978-1-60876-987-2

Bioencapsulation in Silica-Based Nanoporous Sol-Gel Glasses
Bouzid Menaa, Farid Menaa, Carla Aiolfi-Guimarães and Olga Sharts
2010. ISBN: 978-1-60876-989-6

ZnO Nanostructures Deposited by Laser Ablation
M. Martino, D. Valerini, A.P. Caricato A. Cretí, M. Lomascolo and R. Rella
2010. ISBN: 978-1-61668-034-3

Synthesis and Engineering of Nanostructures by Energetic Ions
Devesh Kumar Avasthi and Jean Claude Pivin (Editors)
2010. ISBN: 978-1-61668-209-5

Development and Application of Nanofiber Materials
Shou-Cang Shen, Wai-Kiong Ng, Pui-Shan Chow and Reginald B.H. Tan
2010. ISBN: 978-1-61668-931-5
2010. ISBN: 978-1-61668-829-5
(E-book)

Polymers as Natural Composites
Abdulakh K. Mikitaev; Georgii V. Kozlov and Gennady E. Zaikov (Editors)
2010. ISBN: 978-1-61668-168-5
2010. ISBN: 978-1-61668-886-8
(E-book)

Biocompatible Nanomaterials: Synthesis, Characterization and Applications
S. Ashok Kumar, Sea-Fue Wang and Soundappan Thiagarajan (Editors)
2010. ISBN: 978-1-61668-677-2
2010. ISBN: 978-1-61728-078-8
(E-book)

From Gold Nano-Particles Through Nano-Wire to Gold Nano-Layers
V. Švorčík, Z. Kolská, P. Slepička and V. Hnatowicz
2010. ISBN: 978-1-61668-316-0
2010. ISBN: 978-1-61668-722-9
(E-book)

Phase Mixture Models for the Properties of Nanoceramics
Willi Pabst and Eva Gregorova
2010. ISBN: 978-1-61668-673-4
2010. ISBN: 978-1-61668-898-1
(E-book)

Applications of Electrospun Nanofiber Membranes for Bio-separations
Todd J. Menkhaus, Lifeng Zhang and Hao Fong
2010. ISBN: 978-1-60876-782-3

Nanostructured Materials: Classification, Properties and Fabrication
Anees A. Ansari, M. Naziruddin Khan, M. Alhoshan, A.S. Aldwayyan and M.S. Alsalhi
2010. ISBN: 978-1-61668-763-2
2010. ISBN: 978-1-61728-474-8
(E-book)

Low-K Nanoporous Interdielectrics: Materials, Thin Film Fabrications, Structures and Properties
Moonhor Ree, Jinhwan Yoon, Kyuyoung Heo
2010. ISBN: 978-1-61668-749-6
2010. ISBN: 978-1-61728-318-5
(E-book)

Ba(Ti,Zr)O3 – Functional Materials: From Nanopowders to Bulk Ceramics
Adelina Ianculescu and Liliana Mitoseriu
2010. ISBN: 978-1-61668-752-6
2010. ISBN: 978-1-61728-253-9
(E-book)

Gold Nanoparticles as an Antigen Carrier and an Adjuvant
L. A. Dykman, S. A. Staroverov, V. A. Bogatyrev and S. Yu. Shchyogolev
2010. ISBN: 978-1-61668-771-7
2010. ISBN: 978-1-61728-459-5
(E-book)

Energetics and Percolation Properties of Hydrophobic Nanoporous Media
V. N. Tronin and V. D. Borman
2010. ISBN: 978-1-61668-866-0
2010. ISBN: 978-1-61728-461-8
(E-book)

NANOTECHNOLOGY SCIENCE AND TECHNOLOGY

TiO₂ NANOCRYSTALS

SYNTHESIS AND ENHANCED FUNCTIONALITY

JI-GUANG LI
XIAODONG LI
AND
XUDONG SUN

Nova Science Publishers, Inc.
New York

Copyright © 2010 by Nova Science Publishers, Inc.

All rights reserved. No part of this book may be reproduced, stored in a retrieval system or transmitted in any form or by any means: electronic, electrostatic, magnetic, tape, mechanical photocopying, recording or otherwise without the written permission of the Publisher.
For permission to use material from this book please contact us:
Telephone 631-231-7269; Fax 631-231-8175
Web Site: http://www.novapublishers.com

NOTICE TO THE READER

The Publisher has taken reasonable care in the preparation of this book, but makes no expressed or implied warranty of any kind and assumes no responsibility for any errors or omissions. No liability is assumed for incidental or consequential damages in connection with or arising out of information contained in this book. The Publisher shall not be liable for any special, consequential, or exemplary damages resulting, in whole or in part, from the readers' use of, or reliance upon, this material.

Independent verification should be sought for any data, advice or recommendations contained in this book. In addition, no responsibility is assumed by the publisher for any injury and/or damage to persons or property arising from any methods, products, instructions, ideas or otherwise contained in this publication.

This publication is designed to provide accurate and authoritative information with regard to the subject matter covered herein. It is sold with the clear understanding that the Publisher is not engaged in rendering legal or any other professional services. If legal or any other expert assistance is required, the services of a competent person should be sought. FROM A DECLARATION OF PARTICIPANTS JOINTLY ADOPTED BY A COMMITTEE OF THE AMERICAN BAR ASSOCIATION AND A COMMITTEE OF PUBLISHERS.

LIBRARY OF CONGRESS CATALOGING-IN-PUBLICATION DATA
Li, Ji-Guang.
  TiO2 nanocrystals : synthesis and enhanced functionality / authors,
Ji-Guang Li, Xiaodong Li, Xudong Sun.
    p. cm.
  Includes bibliographical references and index.
  ISBN 978-1-60876-838-7 (softcover)
  1. Titanium dioxide crystals. 2. Nanocrystals. I. Li, Xiaodong, 1949-
II. Sun, Xudong, 1951- III. Title.
  QD181.T6L523 2010   620'.5--dc22
                                                              2009051303

Published by Nova Science Publishers, Inc. † New York

CONTENTS

Preface		xi
Chapter 1	Introduction	1
Chapter 2	Controlled Hydrothermal Processing of TiO_2 Nanocrystals	5
Chapter 3	Morphology Engineering of Rutile Nanocrystals	15
Chapter 4	Solution Processing of TIO_2 under Ambient Conditions	25
Chapter 5	Single Molecular Design of Non-metal Doped TiO_2 for Enhanced Photocatalysis	39
Chapter 6	Efficient Doping of TiO_2 Nanocrystals via Radio-Frequency (RF) Thermal Plasma Processing	49
Chapter 7	Conclusion	81
Acknowledgment		83
References		85
Index		89

PREFACE

This chapter discusses phase selective and morphology controllable processing of titania (TiO_2) nanocrystals and their enhanced functionality via doping for photocatalytic, luminescent, and magnetic applications. Materials synthesis has been achieved by two "extreme" techniques: low temperature soft-chemical processing (typically <250°C) and high temperature RF (Radio frequency) thermal plasma processing (typically above 10,000°C). It is demonstrated that anatase, rutile and even the brookite polymorph of TiO_2 can all be selectively synthesized in a phase-pure form via solution processing, sometimes even under near atmospheric conditions. Processing factors that govern phase structure and crystal morphology of the products are discussed. Detailed characterizations are given to the brookite phase with regard to its physicochemical properties and its irreversible phase transition to rutile. It is also shown that quasi-equiaxed rutile nanocrystals, a morphological form rather difficult to make, can be obtained from Degussa P25 via a hydrothermal crystallization/phase transformation process. A single molecular strategy is demonstrated for one-step doping of TiO_2 nanocrystals with non-metallic elements for improved photocatalytic performances. Meanwhile, effective doping of TiO_2 with transition (Co^{2+}) and lanthanide ions (Eu^{3+}, Er^{3+}) has been achieved via thermal plasma oxidation of liquid precursors containing the component elements, and the thus-made TiO_2 nanomaterials exhibit interesting structural features and/or novel functionalities.

Chapter 1

INTRODUCTION

Titanium dioxide (TiO$_2$) nanomaterials currently find wide technological applications including pigments, cosmetics, ultrathin capacitors, photovoltaic cells, and photocatalysis. The compound has three most commonly encountered crystalline polymorphs: anatase, brookite, and rutile. All the three crystal structures are made up of distorted TiO$_6$ octahedra, but in different ways. The rutile phase adopts a tetragonal structure (space group: $D_{4h}^{14}(P4_2/mnm)$, Z =2), in which two opposing edges of each octahedron are shared to form linear chains along the [001] direction and the TiO$_6$ chains are then linked together via corner connection (Figure 1a). Anatase (tetragonal, ($D_{4h}^{19}(I4_1/amd)$, Z=4) has no corner sharing but has four edges shared per octahedron, and its crystal structure can be viewed as zigzag chains of the octahedra linked together through edge sharing (Figure 1b). As for brookite (orthorhombic, $D_{2h}^{15}(pbca)$, Z =8), the octahedra share three edges and also corners, and the dominant structural feature of brookite is a chain of edge sharing: the distorted TiO$_6$ octahedra are arranged parallel to the c axis and are cross-linked by shared edges (Figure 1c). The crystal built-up (in terms of the number of shared edges) and some known physical properties of brookite seem to go between those of anatase and rutile. For example the refractive index of anatase, brookite, and rutile increases in the order 2.52, 2.63, and 2.72, while the theoretical density in the order 3.84, 4.11, and 4.26 g/cm^3.

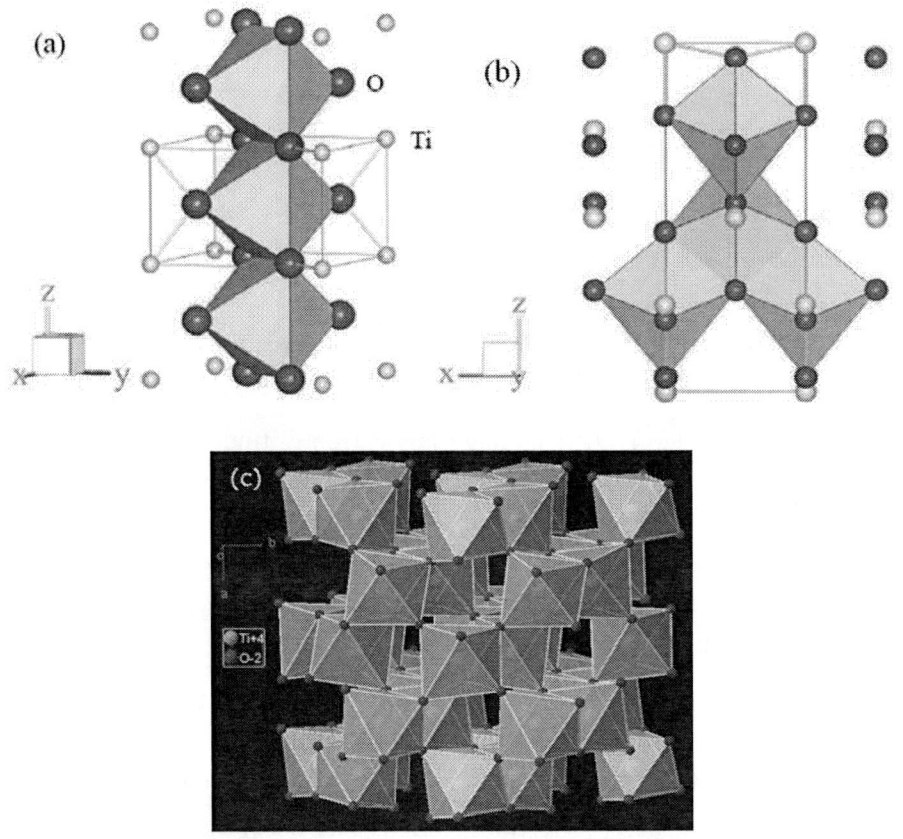

Figure 1. Structure models of rutile (a), anatase (b), and brookite (c) polymorphs of TiO$_2$. The crystal structures of anatase (b) and brookite (c) are viewed along [010] and approximate [001] directions, respectively.

Owing to these different structural features, the three polymorphs of TiO$_2$ exhibit their own characteristic Raman scatterings. From the irreducible presentation of the optical modes, the three phase of anatase, brookite, and rutile have 6 ($3E_g + 2B_{1g} + A_{1g}$), 36 ($9A_{1g} + 9B_{1g} + 9B_{2g} + 9B_{3g}$), and 4 ($A_{1g} + B_{1g} + B_{2g} + E_g$) Raman active modes, respectively [1-3]. Brookite, either natural or synthetic, shows strong Raman peaks at about 128 (A_{1g}), 153 (A_{1g}),

247 (A_{1g}), 322 (B_{1g}), 366 (B_{2g}) and 636 cm^{-1} (A_{1g}). Anatase exhibits characteristic scatterings at 146 (E_g), 396 (B_{1g}), 515 (A_{1g}) and 641 cm^{-1} (E_g), while rutile gives typical scatterings at 143 (B_{1g}), 235 (two-phonon scattering), 447 (E_g) and 612 cm^{-1} (A_{1g}). Raman spectroscopy, besides its extensive use in phase identification of TiO$_2$, serves as an efficient tool for probing oxygen deficiency of the TiO$_2$ lattice. It is widely observed that increased contents of oxygen vacancies in the crystal structure would lead to increased wavenumber of the anatase E_g mode (146 cm^{-1}) while decreased wavenumber of the rutile E_g mode (447 cm^{-1}). Calibration curves of the Raman spectrum of nanophase anatase and rutile to the material's oxygen stoichiometry has been reported by Parker et al [4]. A similar correlation, however, has not been established for the brookite polymorph.

The practical application of TiO$_2$ strongly depends upon the crystal structure, morphology, and size of the crystallites. Each crystalline modification of TiO$_2$ has different physiochemical properties, such as density, refractive index, and photochemical reactivity. Rutile has the highest density and refractive index among the three phases and therefore has been widely employed in pigments and cosmetics industries. Anatase generally shows better performances than its rutile counterpart in photocatalytic applications, likely due to the longer lifetime of the photo-excited h^+/e^- carriers in the anatase lattice. The brookite phase is the least studied in many aspects of its properties, mainly owing to the difficulties encountered in obtaining its pure form. Size, shape, and phase structure controlled synthesis of TiO$_2$ nanocrystals has long been one of the main themes in TiO$_2$ research, and an excellent account of TiO$_2$ nanomaterials has recently been given by Chen and Mao [5] with regard to synthesis, property, modification, and application. This book mainly focuses on phase selective and morphology controlled processing of the three polymorphs of TiO$_2$ (particularly brookite) and the enhanced functionalities of the nanomaterials via doping with either non-metallic or metallic elements for photocatalytic, photoluminescent, and magnetic applications.

Chapter 2

CONTROLLED HYDROTHERMAL PROCESSING OF TiO$_2$ NANOCRYSTALS

2.1. SELECTIVE SYNTHESIS OF ANATASE, BROOKITE, AND RUTILE

Various synthetic techniques have been utilized in the preparation of TiO$_2$ nanocrystals, among which hydrothermal treatment has been drawing much attention considering that it directly produces well-crystallized nanocrystallites of a wide range of compositions within a short period of reaction time. With alkoxides or tetrachloride as the titanium source, it was showed that selective crystallization of anatase and rutile is readily achievable and the phase selection depends upon several factors including solution pH, reactant concentration, and the mineralizer used [6]. It is widely observed that a highly acidic condition or mineralizers such as NaCl, NH$_4$Cl, and SnCl$_4$ etc. favors rutile, while the presence of some carboxylic acids promotes anatase crystallization [6]. Albeit brookite is frequently encountered as a byproduct in the sol-gel or hydrothermal products, there is no doubt to say that this phase is the most difficult to obtain as nanocrystallites in a phase-pure form. It is also worth noting that most of the previous studies on TiO$_2$ synthesis started with titanium (□) compounds, typically TiCl$_4$ and alkoxides, which are both highly sensitive to atmospheric moisture and therefore require special precautions. Titanium trichloride ($TiCl_3$) solution, which shows better stability and is easily manipulatable, may serve as a good source for TiO_2 synthesis [7,8]. The generation of $Ti^{IV}O_2$ from Ti^{III} ions requires an oxidation reaction, which can be achieved with an additional oxidant, such as hydrogen peroxide (H_2O_2), ammonium peroxodisulfate (($NH_4)_2S_2O_8$), nitric acid (HNO_3), perchloric acid ($HClO_4$), or even atmospheric oxygen. Phase structure and crystal morphology of the products autoclaved (180 °C, 3h) from mixed aqueous solutions containing $TiCl_3$, additional oxidant, and sometimes urea (for pH adjustment) are heavily dependent upon the type of oxidant, solute concentration, and solution pH [8]. Keeping the $TiCl_3$/oxidant molar ratio at unity, a systematic study on the effects of processing parameters on characteristics of the hydrothermal products indicate that the use of $(NH_4)_2S_2O_8$ consistently yields anatase, irrespective of the other synthetic conditions (solution pH and the initial Ti^{3+} concentration), though a higher Ti^{3+} concentration yields aggregated anatase crystallites while a lower concentration tends to produce dispersed ones. With the other processing parameters fixed, the use of $HClO_4$, H_2O_2, or HNO_3 as the oxidant produces almost identical results, and in this case phase selection of anatase, brookite, and rutile nanocrystallites can be attained by controlling the reactant ($TiCl_3$) concentration and the solution pH. Some typical morphologies of the resultant TiO_2 nanocrystals are shown in Figure 2. It can be seen that the three polymorphs have their distinctive crystal shapes: nanospheres for anatase, nanorods for rutile, and nanoplates for brookite. It is also clear that a lower initial $TiCl_3$ concentration tends to yield more dispersed crystallites of anatase and rutile. Both selected area electron diffraction (SAED) and lattice fringe analysis via HR-TEM suggest that the rutile nanorods are growing along the [001] direction (*c*-axis) of the crystal structure, while surfaces of the brookite nanoplates are terminated with the high energy (111) planes. The brookite nanocrystals obtained in this way possess a significantly higher specific surface area than those synthesized otherwise [9].

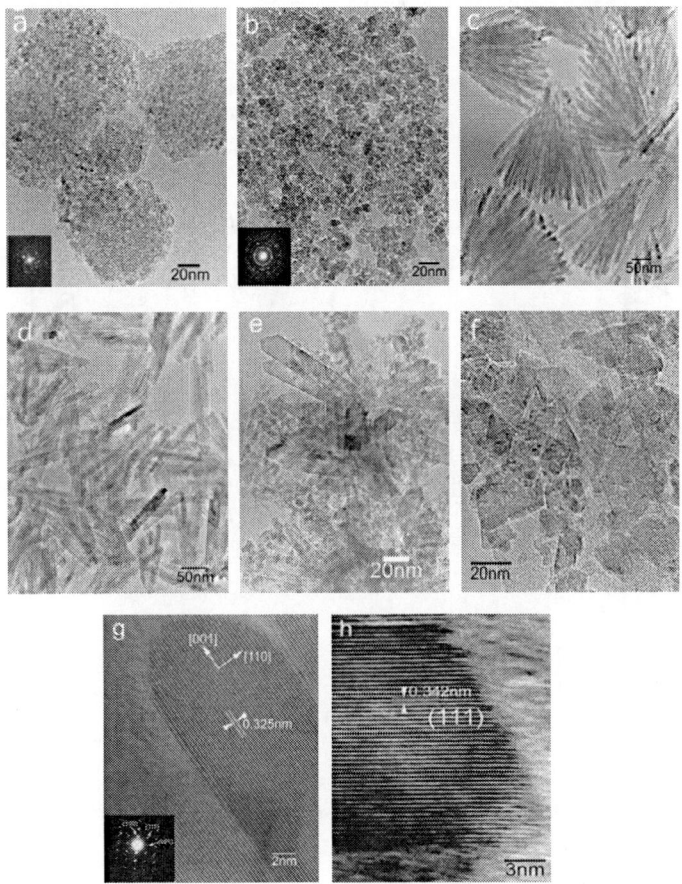

Figure 2. TEM images showing morphologies of the anatase (a, b), rutile (c, d), and brookite (f) nanocrystallites. (e) is a mixture of the three polymorphs of anatase (51.4 wt%, rounded crystallites), brookite (5.0 wt%, platelike), and rutile (43.6 wt%, rodlike). Sample (a) is made with ammonium peroxodisulfate as the oxidant while the others are synthesized with hydrogen peroxide. The synthetic conditions are: $[TiCl_3]$ = 0.9 M, pH < 0 for (a); $[TiCl_3]$ = 0.0625 M and pH = 9.0 for (b); $[TiCl_3]$ = 0.6 M and pH = 0.25 for (c); $[TiCl_3]$ = 0.0625 M and pH = 0.44 for (d); $[TiCl_3]$ = 0.45 M and pH = 0.4 for (e); $[TiCl_3]$ = 0.0625 M and pH = 1.32 for (f). (g) is the lattice image of a single rutile nanorod shown in Figure 2b, from which growth of the nanorod along the [001] direction is determined; (h) is the lattice image of a single brookite nanoplate shown in Figure 2f, where the spacing of 0.342 nm corresponds to the (111) plane of the brookite lattice. Specific surface areas are 297, 187, 47, 67, and 84 m^2/g for powders (a), (b), (c), (d), and (f), respectively.

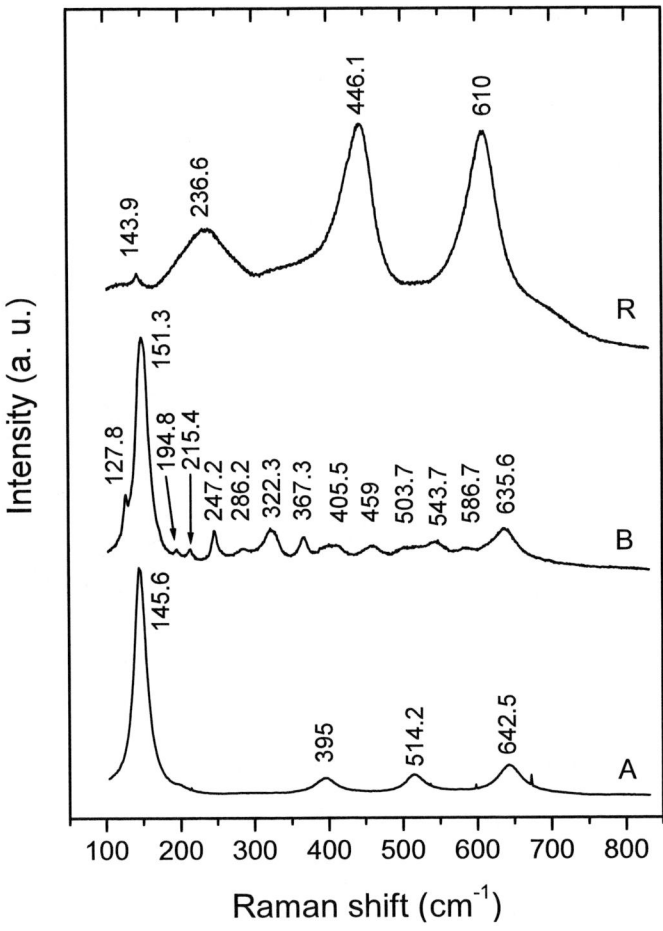

Figure 3. Raman spectra of the three phase-pure TiO_2 powders shown in Figure 2b (anatase), Figure 2d (rutile), and Figure 2f (brookite). A, B, and R denote anatase, brookite, and rutile, respectively.

Phase purity of the anatase, brookite, and rutile nanocrystals are verified via Raman spectroscopy, and the results are displayed in Figure 3. For each phase, both the position and the relative intensity of the observed Raman bands are in good accordance with the literature, indicating high phase purity of these hydrothermally derived TiO_2 nanocrystals. The three phases are also oxygen stoichiometric, though derived from a Ti^{3+} staring precursor, as no shifting of the Raman band owing to the presence of oxygen vacancy is observed. The brookite phase gives additional scatterings besides the strong ones mentioned earlier, whose positions also coincide well with the reported ones, implying high quality of the brookite nanocrystals.

2.2. PHASE SELECTION MECHANISM

It is now the time to address the phase selection mechanism. The results obtained in this work suggest the importance of solution pH, $TiCl_3$ concentration, and supporting anions in the phase selective crystallization of TiO_2. It is generally observed that (1) pure rutile can only be formed under highly acidic conditions, irrespective of the titanium concentration; (2) higher solution pH favors the anatase phase; (3) brookite crystallization is a function of both solution pH and $TiCl_3$ concentration, and a mild pH and an intermediate concentration stabilize the brookite phase. Ti^{4+} cation, oxidized from Ti^{3+} here, generally has a coordination number of 6 and adopts an octahedral symmetry in an aqueous solution. Due to its high charge/radius ratio, Ti^{4+} exhibits a strong tendency to form $[Ti(OH)_x(OH_2)_{6-x}]^{4-x}$ species via hydrolysis in an aqueous solution (here the possible coordination by supporting anions such as Cl^-, SO_4^{2-}, NO_3^- etc was tentatively not considered) [6,10]. Under fixed conditions (reaction temperature and time), the phase selection of TiO_2 would be determined by the interactions among the hydrolyzed precursor cations. According to the "partial charge model" [10], structure of the precursor cation, which is affected by the exact pH, determines the final crystal structure of the resultant TiO_2. The crystallization of TiO_2 starts with $[TiO(OH)(OH_2)_4]^+$, which is evolved from Ti^{4+} via $[Ti(OH)(H_2O)_5]^{3+}$ and $[Ti(OH)_2(H_2O)_4]^{2+}$ species with successively increased solution pH. The $[TiO(OH)(OH_2)_4]^+$ complex can yield $[Ti(OH)_3(OH_2)_3]^+$ via intramolecular deoxolation depending on the exact pH. In the relatively low pH region, deoxolation has much less probability to take place and the oxolation among $[TiO(OH)(OH_2)_4]^+$ results in a linear growth along the equatorial plane of the cations. This reaction, followed by the oxolation between the resultant chains, yields rutile. At high pH, condensation can proceed along apical directions of the $[TiO_6]$ octahedron via deoxolation, leading to the skewed chains of the anatase structure [10]. Pottier et al. [11] suggested that the Cl⁻ ions in the reaction system template the growth of the brookite structure, and indeed the brookite phase frequently appears in the TiO_2 products synthesized in the presence of chlorine ions. Even though, solution pH may still be a decisive factor, since it determines the amount of Cl⁻ in the precursor ions upon condensation (the higher the pH, the fewer the Cl⁻ ions). From another point of view, the brookite structure has both shared edges and corners and is midway between those of anatase and rutile in terms of the shared edges (4 for anatase, 3 for brookite, and 2 for rutile). This might be the reason why brookite needs intermediate pH to stabilize. While these explanations may hold for direct crystallization of the final crystal structure, they do not consider kinetic factors, which would be important especially when the pH is not optimal for direct crystallization (that is, not high enough for anatase or not low enough for rutile). In such a case, phase structure of the final product might be a result of recrystallization (or partial recrystallization) of a transient phase. There have been extensive reports showing that amorphous TiO_2 may transform to anatase, mixtures of anatase and brookite, and pure rutile under certain hydrothermal conditions.

Under a pH suitable for brookite crystallization (pH = 1.32, Figure 2f), it is observed that the titanium concentration affects phase structure of the resultant TiO_2. That is, either increased or decreased concentrations enhance anatase crystallization. The high-concentration effect is believed to be due to the rapid interactions among the precursor cations, preventing sufficient rearrangements of the $[TiO_6]$ octahedrons for brookite structure formation. On the other hand, lower titanium concentration causes decreased growth/nucleation ratio, yielding finer TiO_2 crystallites. It is known that the relative stability of TiO_2 polymorphs is size-dependent: anatase is the most stable at sizes below about 11 nm while brookite in the range about 11-35 nm [12]. The decreased crystallite size destabilizes brookite, leading to the crystallization of anatase. Indeed, the pure brookite (Figure 2f) crystals show overall sizes greater than those of anatase (typically <10 nm, Figure 2b).

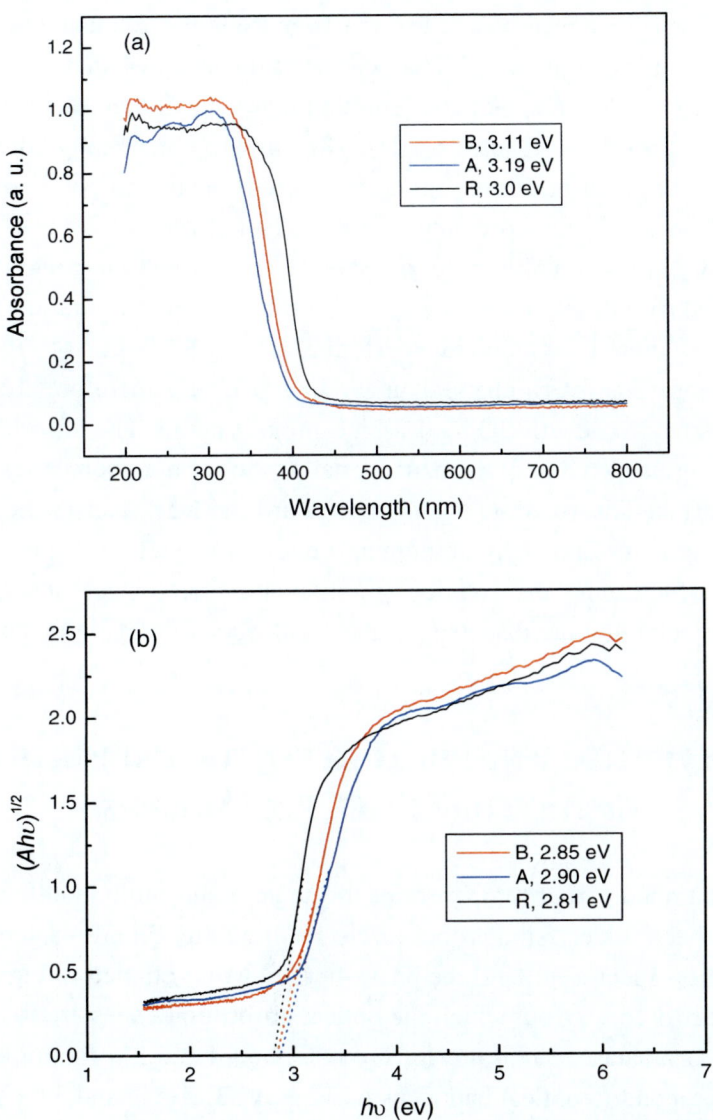

Figure 4. UV-vis absorption spectra (a) and the determination of indirect interband transition energies (b) for the pure anatase, brookite, and rutile powders shown in Figures 2b, 2d, and 2f, respectively. A, B, and R denote anatase, brookite, and rutile, respectively. A in the Y axis title of part (b) represents absorbance, which is proportional to the absorption coefficient α.

The anions generated via redox reaction from the oxidant may act as ligands to complex with Ti^{4+}. It has been well documented that SO_4^{2-} shows higher affinity to Ti^{4+} in an aqueous solution than many other anions do (such as Cl^-, NO_3^-, ClO_3^-, ClO_4^-, and $-OOH$). The strongly chemically coordinated SO_4^{2-} anions seem to orient the titanium precursor cations upon deoxolation and oxolation, and thus promote anatase crystallization. By varying the reaction conditions, another role of SO_4^{2-} is confirmed: it coagulates the anatase nanocrystallites under low pH or high SO_4^{2-} concentration to form wormhole-structured agglomerates (Figure 2a). Surfaces of the TiO_2 crystallites are positively charged under low pH, and therefore they show strong tendency to adsorb SO_4^{2-} of a high minus charge. The adsorbed SO_4^{2-} acts as a coagulant, tying up the nanocrystallites to form agglomerates (Figure 2a). The surface-adsorbed SO_4^{2-} may also retard the mass transfer needed for crystallite growth under hydrothermal conditions, yielding finer anatase crystallites. Increasing the solution pH alleviates the surface adsorption of SO_4^{2-}, thus yielding more dispersed anatase nanocrystallites (Figure 2b).

2.3. OPTICAL AND PHOTOCATALYTIC PROPERTIES OF THE THREE TiO_2 POLYMORPHS

As mentioned earlier, properties of the anatase and rutile modifications of TiO_2 have been widely studied but rarely for brookite. Figure 4a shows UV-vis absorption spectra of the three phase-pure powders studied in Figure 3 via Raman spectroscopy, from which the optical absorption edges are found to be 388.7 nm for anatase, 398.6 nm for brookite, and 413.3 nm for rutile. These values correspond to optical bandgaps of 3.19 eV, 3.11 eV, and 3.0 eV for the three phases, respectively. The optical bandgaps of anatase and rutile determined here are in excellent agreement with the widely observed ones (~3.2 eV for anatase and 3.0 eV for rutile). The optical bandgap of brookite is largely unknown, but the data obtained here indicates that it goes in between those of anatase and rutile. TiO_2 is known as an indirect semiconductor [13], for which the relation between absorption coefficient (α) and incident photon energy (hv) can be written as $a = B_i(hv - E_g)^2 / hv$, where B_i is the absorption constant for indirect transitions. Plots of $(Ahv)^{1/2}$ versus hv from the spectral data of Figure 4a are presented in Figure 4b. Extrapolating the linear part of the curve for anatase gives an indirect bandgap of 2.90 eV, which is in close vicinity to the calculated value of 2.91 eV corresponding to $X_{1a} \rightarrow G_{1b}$ indirect interband transition. The indirect bandgaps of brookite and rutile are similarly determined to be 2.85 eV and 2.81 eV, respectively. Again, the value of brookite falls in between those of anatase and rutile.

Photocatalytic performances of these TiO_2 nanopowders are tested via bleaching 20 µM methyl orange solutions for four typical samples: wormhole-structured anatase (Figure 2a), dispersed anatase nanocrystallites (Figure 2b), rutile nanorods (Figure 2d), and brookite nanoplates (Figure 2f). The catalytic performance of a TiO_2 powder appears to be a delicate function of many factors [13], particularly phase structure, impurity (dopant) type and concentration, specific surface area, crystallite size, crystal shape, and surface hydroxyls. Among these factors, specific surface area plays an important role, as the catalytic reactions proceed through the interactions between the adsorbate and the free h^+/e^- that have survived the migration from the crystallite interior to surfaces. To exclude the effects of specific surface area, surface areas of all the tested powders are set the same as 3 m^2 by varying the sample weight. Figure 5 shows degradation kinetics of the methyl orange solution, where C_0 and C represent the original and the instant concentrations of methyl orange aqueous solutions, respectively. Photocatalysis is performed by shining UV light (1 mW/cm^2) on the top surface of 10 ml dye solution (in cylinder bottle of 20 ml capacity) with a certain amount of TiO_2 nanocrystallites ultrasonically dispersed in it. All the photocatalytic reactions are conducted under magnetic stirring of the suspension. After being illuminating for a certain period of time, the suspension is centrifuged under 12,557×g for 30 min to achieve solid/liquid separation, and the relative concentration of methyl orange in the recovered solution is determined through UV/Vis spectroscopy by comparing the intensity of the 465 nm absorption of the recovered solution with that of the original methyl orange solution. A blank test indicates that in the absence of TiO_2 photocatalyst decoloration of the dye solution is negligible within the tested period of 2 h. In the presence of TiO_2 nanocrystallites, UV irradiation results in decoloration in all the cases, suggesting destruction of the absorption band of methyl orange by the TiO_2 photocatalysts. The rutile powder shows a performance inferior to anatase as commonly observed, though it has a lower bandgap (Figure 4) and a higher absorption in the UV region (Figure 4a). The two anatase powders, though differ in overall morphology, exhibit quite similar efficiencies. The best performance is observed for the brookite nanoplates, which causes an almost complete decoloration of the methyl orange solution within 2 h. It is believed that the peculiar plate-like morphology of the brookite nanocrystals (with the high energy (111) facets exposed) and the higher light-absorption of brookite than anatase are responsible for the excellent photocatalytic performance of the brookite nanoplates.

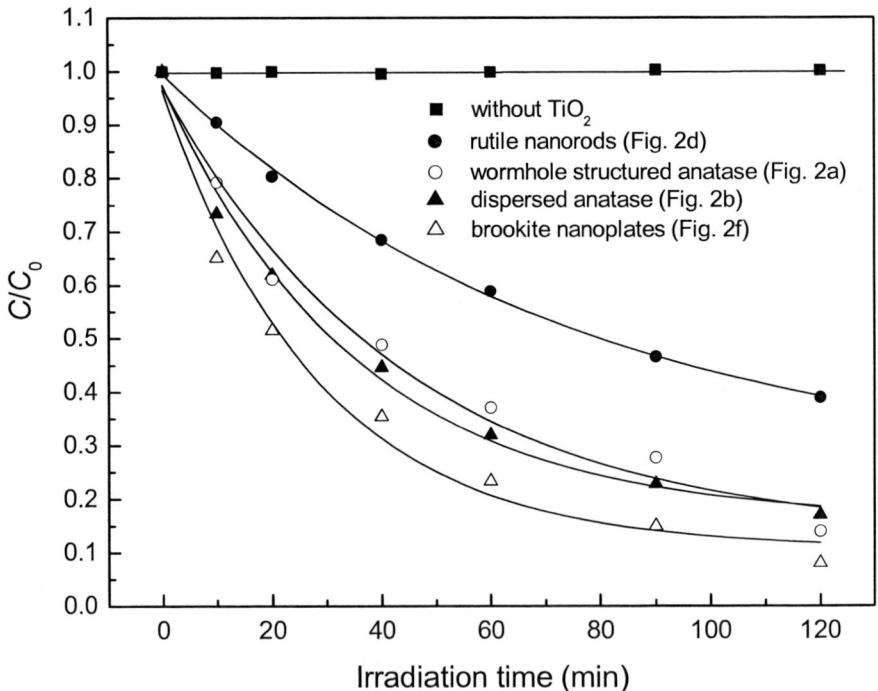

Figure 5. Degradation kinetics of 20 μM methyl orange solutions over TiO_2 photocatalysts under UV irradiation (1 mW/cm^2).

Chapter 3

MORPHOLOGY ENGINEERING OF RUTILE NANOCRYSTALS

It has been a wide observation that the rutile phase hardly crystallizes as equiaxed grains but tends to form nanorods via hydrothermal synthesis or hydrothermal crystallization of amorphous TiO_2, as can also be seen from Figures 2c,d. The reason for this anisotropic growth habit has been proposed to be due to the unique crystal structure of rutile. That is, rutile has 4_2 screw axes along the crystallographic c-axis and this screw structure tends to promote the crystal growth along this direction [14]. Till now, the most successful synthesis of equiaxed rutile nanocrystallites seems to be that reported by Aruna et al [15]. By autoclaving at 250 °C for 26 h under magnetic stirring the reaction products of titanium isopropoxide and nitric acid (pH=0.5), they obtained tetragonal shaped rutile nanocrystals (~22 nm) of good dispersion without using any mineralizer. Li and Ishigaki [16] made equiaxed rutile nanocrystals by $Ar-CO_2-H_2$ thermal plasma oxidation of commercially available TiC particles, but owing to the intentionally introduced reducing gas (H_2, indispensable to pure rutile) the resultant rutile nanopowder appears bluish green due to the pronounced oxygen deficiency. Accordingly, the E_{Rg} band of the resultant rutile significantly red-shifted from 447 cm^{-1} to ~416 cm^{-1}.

PHASE CONVERSION OF DEGUSSA P25 UNDER HYDROTHERMAL

The Degussa P25 type TiO_2 powder has been a focus of studies for years owing to its unique phase structure and particularly its outstanding photocatalytic performance. The powder is also a frequent staring material for further processing. Morgan et al. [17] very recently reported a "morphological phase diagram" for the titania/titanate nanostructures made via alkaline hydrothermal treatment of Degussa P25. Though the phase transition from metastable anatase to thermodynamically stable rutile has been extensively studied via conventional annealing, phase transition of the anatase portion of P25 into rutile and morphologies of the resultant rutile crystallites have rarely been investigated under acidic hydrothermal conditions. We addressed these issues and found that equiaxed rutile nanocrystals can be obtained via hydrothermal (180 °C for 64 h) treating P25 in the presence of nitric acid. The Degussa P25 powder used for this acidic hydrothermal treatment has ~82.1 wt% of anatase and 17.9 wt% of rutile, and the average crystallite sizes of the anatase and rutile phases are assayed from the Scherrer formula to be ~24.7 nm and 37.6 nm, respectively. BET analysis found a specific surface area of ~48.6 m^2/g for this P25 powder. All these data are in general consistence with the widely reported values.

Conversion of the anatase fraction of P25 into rutile involves a dissolution-recrystallization process and acid addition is indispensible. It is found that, without any acid addition, autoclaving an aqueous suspension of P25 at 180 °C for 64 h does not alter appreciably the phase constituent. Preliminary experiments show that this hydrothermal conversion process is rather sluggish due to the high crystallinity of the P25 powder arising from high-temperature flame pyrolysis. At a nitric acid addition of 1.5 mL (HNO_3/TiO_2=1.8 in molar), autoclaving at 180 °C for 7 days only yields a phase mixture containing ~33 wt % of rutile even though pH of the system has been below zero. At a reasonable autoclaving length of 64 h, we studied the effects of HNO_3 addition on the anatase conversion, and the results are given in Figure 6. Clearly, a higher amount of HNO_3 addition significantly enhances the transformation process, and phase-pure rutile can be obtained by adding 10 mL of HNO_3 (HNO_3/TiO_2=12 molar ratio) or more.

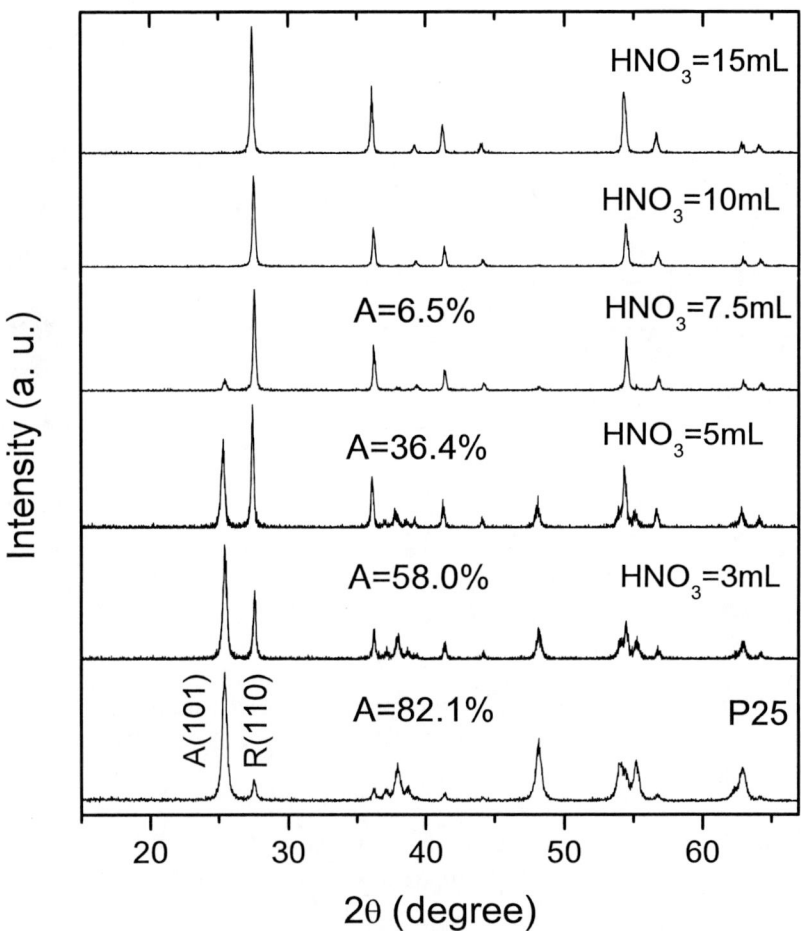

Figure 6. XRD patterns of the hydrothermal products obtained at 180 °C for 64 h, as a function of HNO_3 addition. A and R denote anatase and rutile, respectively.

The hydrothermal conversion process has also been studied at 180 °C by fixing the HNO_3 addition at 10 mL, and XRD patterns of the hydrothermal products are shown in Figure 7. A progressive conversion of the anatase phase into rutile is clearly seen.

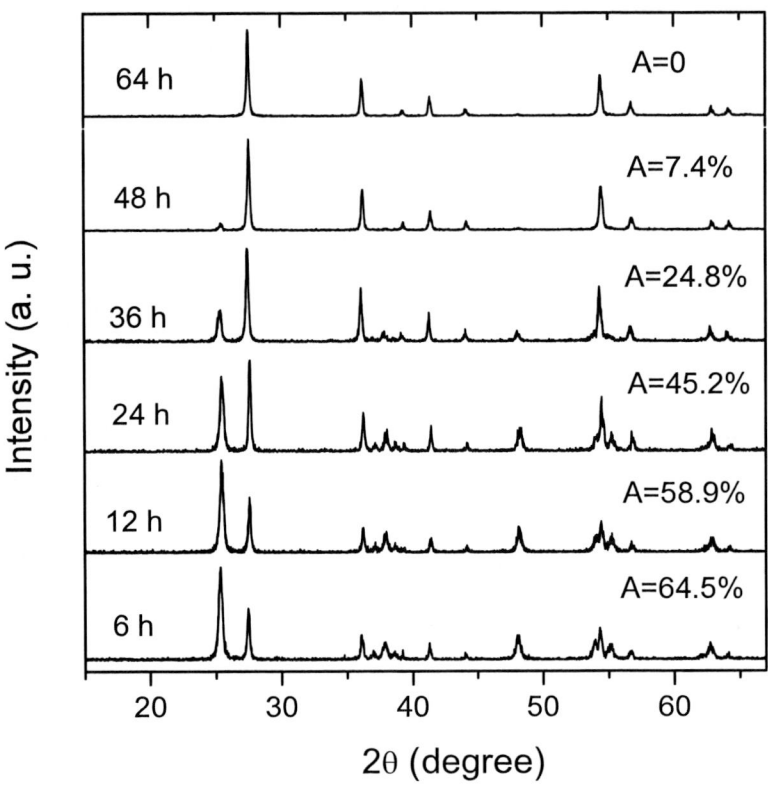

Figure 7. XRD patterns of the hydrothermal products obtained at 180 °C, as a function of treating time. The amount of HNO_3 added has been fixed at 10 mL (HNO_3/TiO_2=12 molar ratio) in each case.

The amount of anatase decreases almost linearly along with an increase in the treating time (Figure 8), suggesting a zero-order reaction of the phase conversion. Fitting the experimental data by linear regression yields a constant rate (slope of the line) of ~1.469 wt%/h for this conversion process. Besides, unlike hydrothermal crystallization of amorphous TiO_2, the brookite polymorph of TiO_2 does not appear here (Figures 6, 7), conforming to the previous observations that a high acidity favors rutile while brookite tends to crystallize under a mild pH.

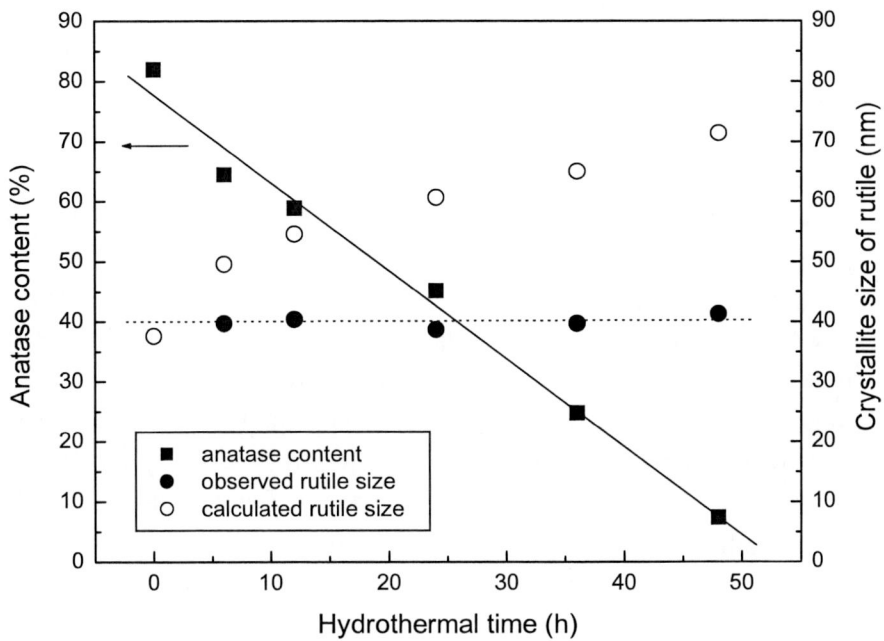

Figure 8. Anatase content (filled rectangles) and experimentally determined rutile size (filled circles) of the hydrothermal products, as a function of the treating time. Treating temperature and the amount of HNO_3 added have been kept constant at 180 °C and 10 mL, respectively. Open circles represent the crystallite sizes calculated according to the "seeded transformation and epitaxial growth" mechanism.

Another interesting finding is that the average crystallite size of rutile remains almost constant irrespective of the hydrothermal time (Figure 8, filled circles). This is somewhat out of expectation, as it was expected that the ~18 wt% of rutile crystallites in the starting P25 powder might be able to seed the phase conversion and thus lead to epitaxial growth of the existing rutile crystallites. Assuming instantaneous site saturation followed by steady spherical growth and that the rutile crystallites are identical spheres, the rutile size evolved via "seeded transformation and epitaxial growth" might be estimated, and the results are summarized in Figure 8 (open circles). Clearly, the rutile sizes predicted from "seeded growth" are constantly much larger than the experimentally observed ones, indicating that the "seeded transformation and epitaxial growth" mechanism hardly occurs to a significant extent, though it can not be completely excluded. The nearly constant mean crystallite size of rutile observed from Figure 8 might be understood from the following: (1) the nucleation of rutile proceeds throughout the whole process of hydrothermal treatment due to the rather low solubility product of TiO_2, and the smaller crystallites (especially the newly formed nuclei) kinetically grow faster than the bigger ones. That is, the titanium ionic species generated via anatase dissolution under hydrothermal are largely consumed in the growth of the finer rutile crystallites. As the rutile size shown in Figure 8 is averaged from the finer crystallites and the bigger ones, it is possible that the mean crystallite size remains almost constant against the hydrothermal time; (2) rutile turns thermodynamically stable when its crystallite size reaches ~35 nm [12], and hence finer crystallites have a strong tendency to reach this value for stabilization. The combined effects of the above two may thus have yielded an average crystallite size of ~41 nm, a value not far from the 35-nm requirement for rutile stabilization and also not far from the original rutile size of ~38 nm in the starting P25 powder.

Figure 9 shows TEM images of the hydrothermal products obtained under different conditions. It is observed that the pure rutile nanocrystallites obtained with 10 mL of HNO_3 addition (HNO_3/Ti=12) largely assume tetragonal morphologies and are quasi-equiaxed (Figure 9a). Average crystallite size of this powder was assayed from the Scherrer equation to be ~41.3 nm. Crystallites with sizes greater than ~70 nm are rarely observed from Figure 9a, further confirming the less probability of the "seeded transformation and epitaxial growth" mechanism.

Increasing the HNO_3 addition to 15 mL (HNO_3/Ti=18), however, yields rutile nanocrystallites of significantly elongated shapes (Figure 9b), suggesting that under this condition of even higher acidity the thermodynamically stable rutile has undergone dissolution-reprecipitation, at least partially. Most of the rodlike particles are observed to have diameters of ~30-50 nm and lengths of up to ~300 nm. As mentioned earlier, these nanorods are growing along the [001] direction (c-axis) of the rutile structure. Replacing HNO_3 with HCl while keeping the acid/TiO_2 molar ratio at 12 similarly yields rutile nanorods by autoclaving at 180 °C for 64 h, but the rods turn much thinner (up to ~20 nm) in diameter and exhibit significantly higher aspect ratios (Figure 9c). Such a phenomenon indicates that Cl⁻ appreciably promotes the development of rodlike morphologies. BET analysis found specific surface areas of ~31.5, 21.8, and 24.7 m^2/g for the rutile powders shown in parts a, b, and c of Figure 9, respectively.

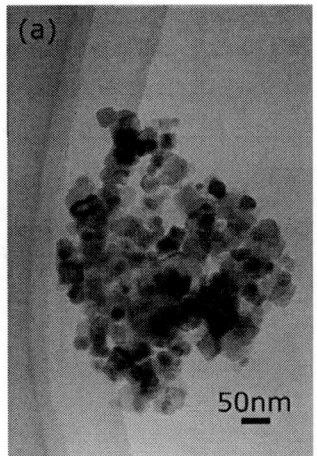

Figure 9. TEM images showing morphologies of the hydrothermal products obtained at 180 °C for 64 h, with (a) HNO_3/TiO_2=12, (b) HNO_3/TiO_2=18, and (c) HCl/TiO_2=12 molar ratio.

3.2. SOLUTION CHEMISTRY AFFECTING RUTILE MORPHOLOGY

The conversion of metastable anatase into rutile involves a dissolution-reprecipitation process as aforementioned. That is, anatase is progressively dissolved by the nitric acid under hydrothermal conditions to generate ionic species of titanium, which then re-precipitate as thermodynamically stable rutile upon the solubility product of TiO_2 is reached. The much finer average crystallite size of anatase than that of rutile in the starting P25 powder would favor this preferential dissolution process. Thermodynamically stable rutile can also be dissolved under hydrothermal, but only under more acidic conditions (Figure 9b). Titanium ions adopt octahedral coordination symmetries in an aqueous solution and they might be generally expressed as $[TiO_z(OH)_x(OH_2)_y]_m^n$ ($0 \leq z \leq 1$, $x+y+z=6$), where the number of x depends on the exact solution pH (the lower the pH, the smaller the x value). Under the highly acidic conditions employed in this work (pH<0), the ionic species generated via dissolution are most likely $[TiO(OH)(OH_2)_4]_m^n$ [15]. The formation of rutile structure under acidic conditions has been discussed earlier according to the "partial charge" model. That is, the oxolation (the only possible reaction under highly acidic conditions) among these ionic species leads to a linear growth of the [TiO_6] building blocks along the equatorial plane of the titanium cation, and this reaction, followed by the oxolation among the resulting linear chains, yields rutile crystals. This formation mechanism conforms to the crystal structure of rutile: two opposing edges of each [TiO_6] octahedron are shared to form linear chains along the [001] direction and the [TiO_6] chains are then linked together via corner connection. It is perceivable from the above mechanism that morphologies of the resultant rutile crystallites would depend upon the extent of oxolation among the linear chains, and a less extent would favor rodlike morphologies. With this understanding, it can be inferred from Figure 9 that a higher acidity or the existence of Cl⁻ appreciably blocks the oxolation reactions among the linear chains. The effects of Cl⁻ are more pronounced, as seen by comparing Figure 9b and Figure 9c, and this is due to the high compexing capability of Cl⁻

toward titanium ions to form $[TiO(OH)(OH_2)_x Cl_y]_m^n$ ionic species (x+y=4) [6]. The Cl⁻ ions coordinated to Ti⁴⁺ may significantly disturb (retard) the oxolation reactions mentioned above. It can thus be concluded from the above arguments that a proper solution pH and especially a Cl⁻-free condition would benefit the crystallization of equiaxed rutile crystallites.

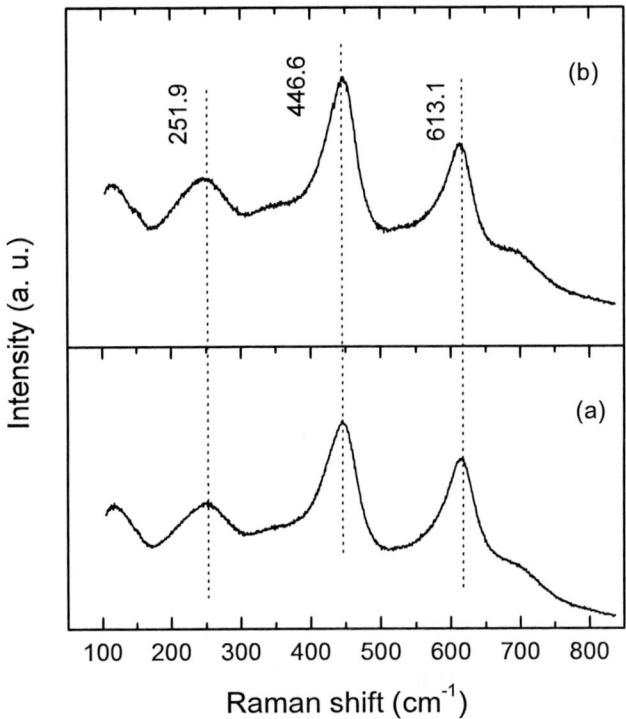

Figure 10. Raman spectra of (a) equiaxed rutile nanocrystals (Figure 9a) and (b) rutile nanorods (Figure 9c).

Figure 10 shows Raman spectra of the hydrothermal products for two typical samples of the tetragonal crystallites (Figure 9a) and the nanorods (Figure 9c). No scattering band corresponding to the anatase polymorph is observed, indicating that they are of pure-rutile. Besides, both the positions and the relative intensities of the resolved Raman bands agree well with those of the stoichiometric rutile. Both the powders appear pure-white due to the lack of oxygen deficiency, which in combination with the high refractive index of rutile, may allow their potential applications as nano-whiteners in cosmetics and pigments.

Chapter 4

SOLUTION PROCESSING OF TIO$_2$ UNDER AMBIENT CONDITIONS

The foregoing chapters discuss the hydrothermal processing of TiO$_2$ nanocrystals of desired phase structure and particle morphologies. As for the mechanism of phase selection, a main conclusion is that the crystallization of anatase, brookite and rutile is dependent upon both solution chemistry and reaction kinetics. A high solution pH and a faster chemical reaction would generally favor anatase, while a strongly acidic condition (therefore slow reaction) favors rutile. The brookite polymorph tends to crystallize under a mild pH and an intermediate reaction rate. With these as a guide, it is shown that the phase structure control of TiO$_2$ can even be achieved under near atmospheric conditions.

MONODISPERSED BROOKITE SPHERES FROM TITANIUM TRICHLORIDE

With titanium trichloride (TiCl$_3$) solution as the titanium source and by reacting a mixed aqueous solution containing TiCl$_3$ (0.015 mol/L) and urea (0.5 mol/L, for pH adjustment) at 70 °C for 2 h, phase pure brookite is resulted under ambient pressure [7]. Further studies indicate that the brookite phase is thermally stable up to ~500 °C, and above which a phase transition to thermodynamically stable rutile is observed (Figure 11). FE-SEM observation (Figure 12a) reveals that the brookite particles are quasi monodispersed and assume a rounded morphology. TEM analysis (Figure 12b) indicates that these spherical particles are stacks of nanoplates of brookite primary crystals (the platelike shape is similar to that shown in Figure 2f). Average size of the brookite cluster is ~154 nm, as determined from the diameters of randomly selected 200 particles. Particle morphology remains nearly intact by annealing up to 500 °C (Figure 12c), but significant densification of the clusters occurs at 700 °C, resulting in polyhedral rutile particles with a smaller average diameter of ~113 nm (Figure 12d). The densification is largely resulted from sintering of the primary brookite crystallites within each cluster at elevated temperatures and may also be due to the brookite → rutile phase transition itself, as rutile has a higher theoretical density (4.26 g/cm^3) than brookite (4.11 g/cm^3). The as-synthesized brookite powder has a specific surface area of ~41.2 m^2/g, which slightly decreases to ~37.8 m^2/g at 500 °C and rapidly decreases to ~9.7 m^2/g after transforming to rutile at 700 °C. The anatase polymorph does not seem to be involved in the brookite → rutile phase transition, as perceived from Figure 11, and this is further confirmed by the Raman spectroscopy studies shown later. It is also conceivable from Figure 12a and Figure 12d that the phase transition mainly takes place within the individual brookite particles (clusters).

The rounded brookite particles (Figures 12a, b) are apparently formed via an aggregation mechanism, and a kinetic model of which has been proposed recently [18]. Similar to the hydrothermal cases, the formation of the brookite polymorph of TiO_2 from a Ti(III) starting precursor needs an oxidation reaction, which is attained in this case via the redox between Ti^{3+} and the atmospheric oxygen ($4Ti^{3+} + O_2 \rightarrow 4Ti^{4+} + 2O^{2-}$). The thus-generated Ti^{4+} ions would undergo subsequent hydrolysis to generate TiO_2 nanoparticles via nucleation/growth. The submicron spheres of Figure 12a are then formed by clustering and the growth of the clusters via diffusion of the primary nanoparticles. Due to the strong Van der Waals forces between a large cluster and its around primary nanoparticles, the latter selectively aggregate to the cluster and then a concentration slope would be created around the cluster. The existence of such a concentration slope would enable the cluster to gather even considerably distant primary nanoparticles, finally yielding the spheres of submicron sizes (Figure 12a).

The direct crystallization of brookite under near ambient condition and without the aid of any mineralizer is made possible by at least two reasons: (1) the slow oxidation of Ti^{3+} by the atmospheric oxygen. The oxidation largely takes place on the surface of the mixed solution of $TiCl_3$ and urea, where the Ti^{3+} ions are in direct contact with O_2. The oxidation reaction is relatively slow, and thus brookite crystallization is kinetically favored. The oxidation reaction is believed to be the rate limiting factor for nucleation/growth and is also crucial to the stabilization of brookite; (2) suitable solution pH. The current reaction system has a moderate final pH of ~2.0, which is beneficial to brookite crystallization as discussed earlier.

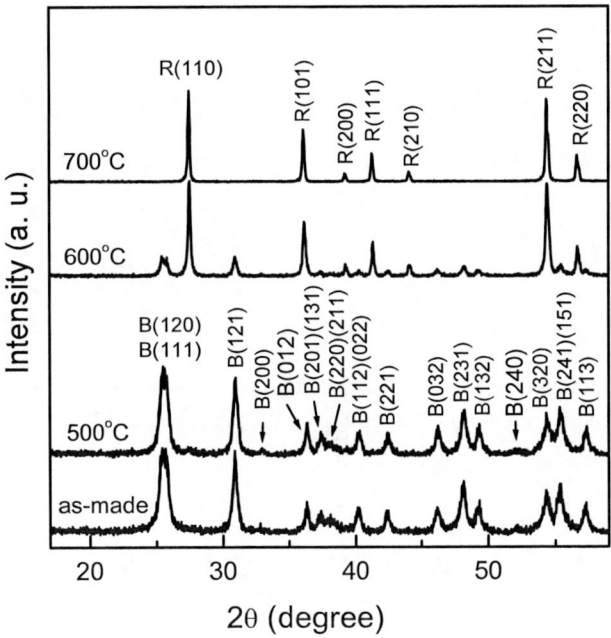

Figure 11. XRD patterns of the as-synthesized brookite powder and its annealing products. The residence time at each annealing temperature is 2 h. B: brookite; R: rutile.

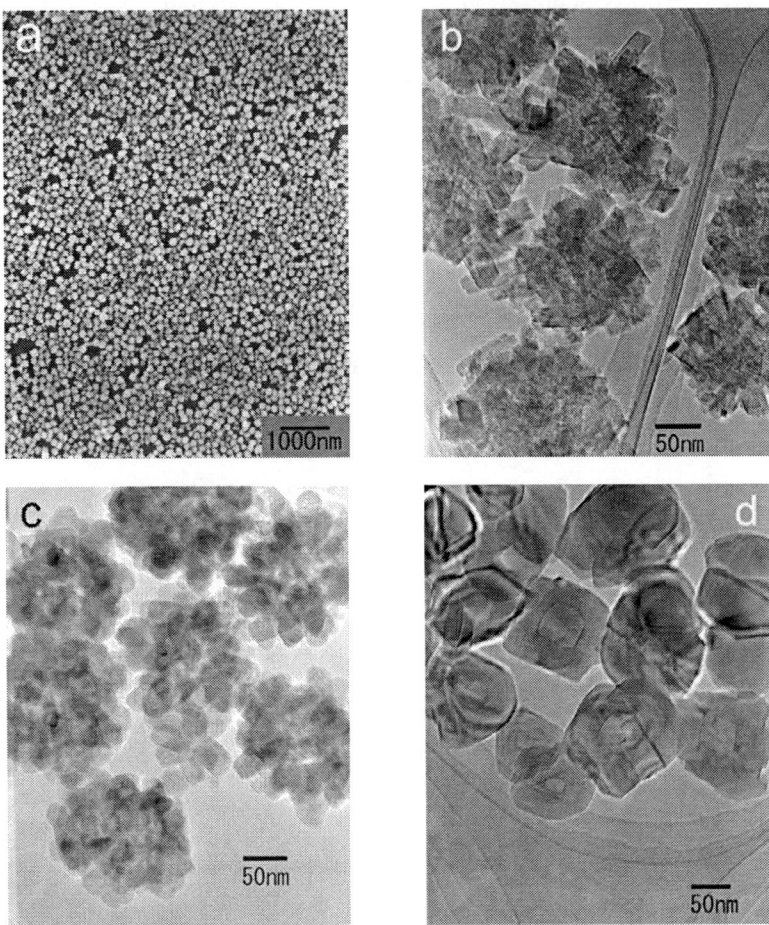

Figure 12. Morphologies of the as synthesized brookite particles (panels a and b) and the products obtained by annealing (a) at 500 (panel c, brookite) and 700 °C (panel d, rutile) for 2 h.

4.2. Phase Transition Phenomena of the Brookite Monospheres

Brookite has been much less characterized as compared to anatase or rutile, mainly due to the difficulties encountered in obtaining its phase-pure form as mentioned earlier. As for phase transformation, the anatase → rutile transition has been studied extensively and a variety of kinetic models have been proposed up to date to interpret the experimental data [19]. It is generally accepted that the transformation takes place in the temperature range ~400-1000 °C depending upon a variety of factors including crystallite size, size distribution, contact area of the crystallites in the powder, impurity type and concentration, and atmosphere. These factors, however, may render uncertainties to the studies of transformation kinetics, which explains why in some cases the experimental data can not be fitted with one single kinetic model and why the derived activation energies are widely dispersed.

The monodispersed quasi-spheres of brookite particles (Figure 12a) provide an ideal starting material for brookite → rutile phase transition studies, since the highly uniform particle morphology may reduce the uncertainties mentioned above. Figure 13 shows XRD patterns of the brookite powder annealed at various conditions. Attentions are paid to the brookite (121) and (120) peaks, as the strongest diffraction from anatase ((101), d=0.352 nm), if there is any, tends to overlap with the strongest diffraction from brookite ((120), d=0.3512), and this may alter the $I_{brookite}^{(121)}/I_{brookite}^{(120)}$ intensity ratio. Ideal brookite has a $I_{brookite}^{(121)}/I_{brookite}^{(120)}$ value of ~0.9, as shown by the diffraction data file (JCPDS: 21-1276). All our samples, the as-made and the partially transformed, have $I_{brookite}^{(121)}/I_{brookite}^{(120)}$ values constant at ~0.9, which may indicate the absence of anatase.

Based upon the results shown in Figure 11, the temperature range 500-600 °C is chosen for phase transformation studies, and the volume fraction (α) of rutile in the powders annealed at various temperatures is shown in Figure 14 as a function of the annealing time. The Johnson-Mehl-Avrami-Kolmogorov (JMAK) model is regarded as universal and has been widely used in the phase transformation and crystallization studies of solids. Though previously the model was successfully employed in the phase transition study of anatase, the application of this model in the present brookite transition, however, only yields smoothly curved lines, indicating a continuously changed Avrami exponent (n) along with the progress of transition. The JMAK model is based upon assumptions that nucleation is a uniform and independent random process and that the parent phase is infinite in size. The non-linearity observed here may mainly arise from this second assumption, as the transformation largely occurs within the individual brookite clusters, which have very limited sizes of ~154 nm.

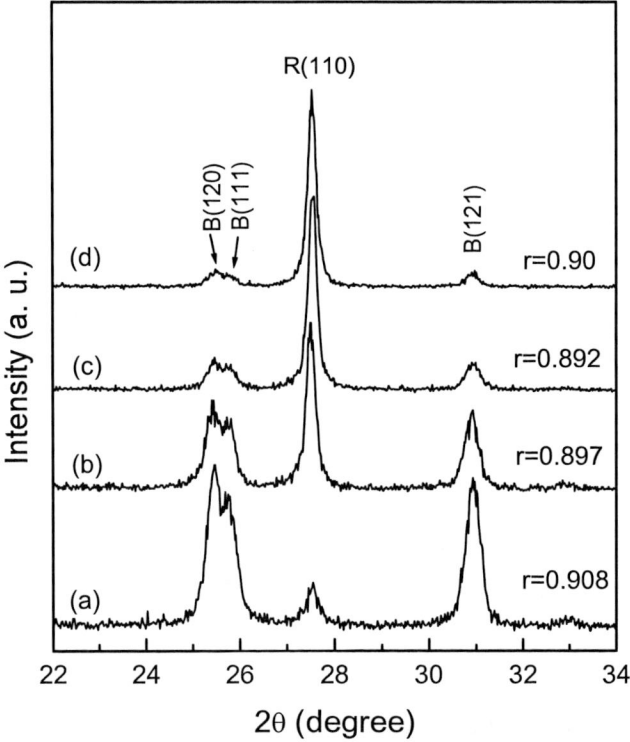

Figure 13. XRD patterns showing relative intensities of the brookite (120) and (121) peaks ($r = I_{brookite}^{(121)} / I_{brookite}^{(120)}$) for the powders annealed to various rutile contents: (a) 5.7 vol% rutile, annealed at 530 °C for 2 h; (b) 34.6% vol% rutile, annealed at 560 °C for 4 h; (c) 68.4 vol% rutile, annealed at 545 °C for 16 h; and (d) 92.1 vol% rutile, annealed at 575 °C for 13 h.

The experimental data shown in Figure 14 are best fitted with the "contracting spherical interface" model, which is based upon spherical particles and the movement of the interface between the parent and new phases inward in the form of a "contracting sphere" [19]. The quasi-spherical brookite clusters used in this work is also morphologically suited to this model:

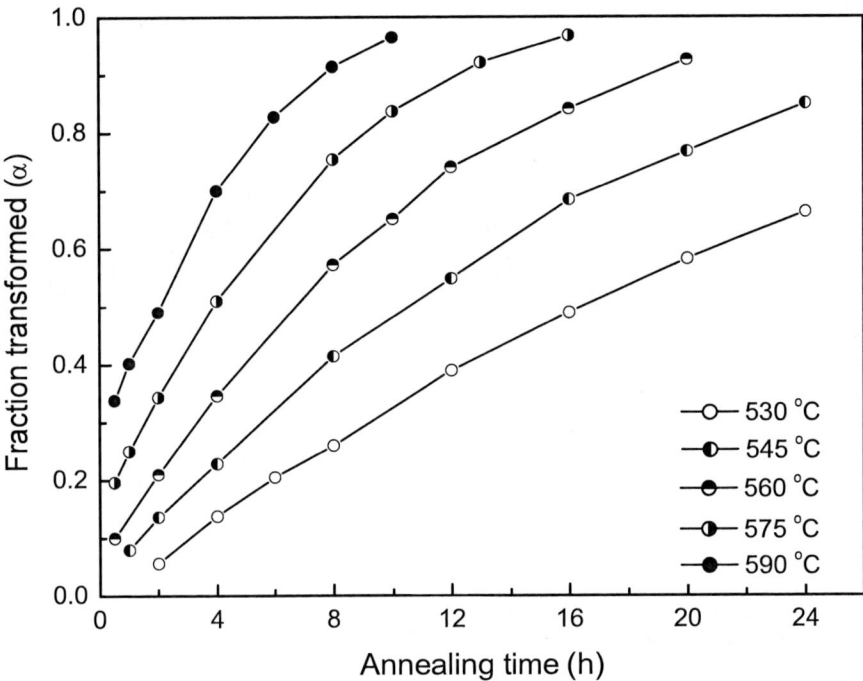

Figure 14. Rutile fractions of the annealed brookite powders, as a function of the annealing temperature and heating time.

where α is the volume fraction of rutile formed within annealing time t, c is a constant, and k is a rate parameter following the Arrhenius law for a thermally activated transition process.

Figure 15 shows the plots of $(1-\alpha)^{1/3}$ against the annealing time t according to eq. (1). Linear relationship holds for the studied annealing temperatures. The Arrhenius plot is shown in Figure 16, from which the apparent activation energy and the pre-exponential factor are determined to be 143.4±1.2 kJ/mol and 1.28×10^4 s^{-1}, respectively.

$$(1-\alpha)^{1/3} = kt + c \tag{1}$$

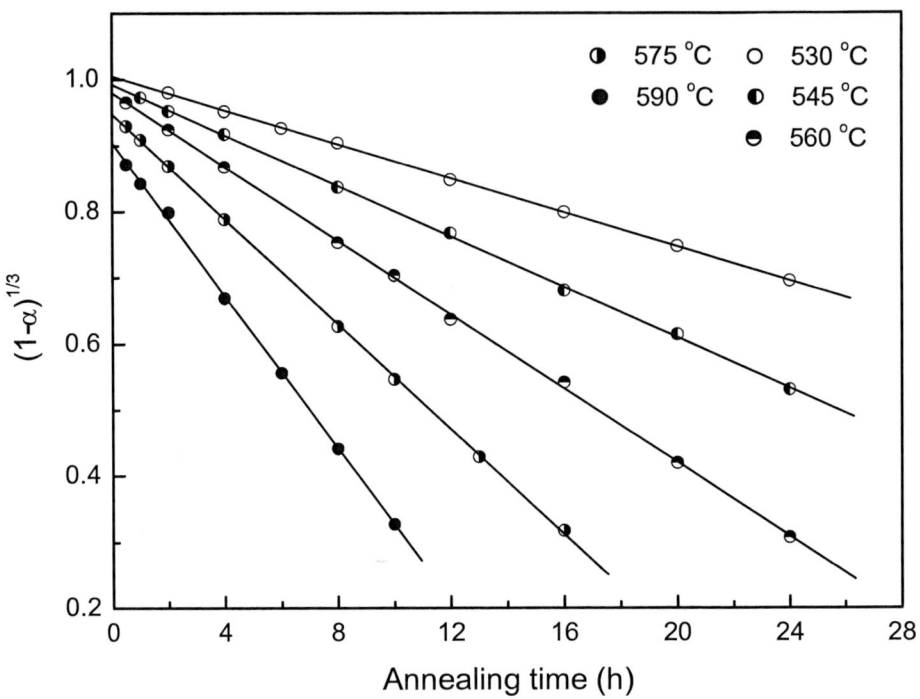

Figure 15. The linear relationship between $(1-\alpha)^{1/3}$ and annealing time t, according to the "contracting spherical interface" model.

No previous experimental work on the brookite transformation is available for a direct comparison. Zhang et al. [12] studied theoretically the brookite → rutile transition in a mixture of anatase (46.7 wt%, ~5.1 nm) and brookite (53.3 wt%, ~8.1 nm) prepared by the sol-gel method. Through enthalpy calculations, they found an apparent activation energy of ~163.8 kJ/mol and a pre-exponential factor of 9.67×10^4 s^{-1}. Our experimental work seems to support their results, as the activation energies are close to each other and the pre-exponential factors are on the same order. Based upon the experimental results, the rate factor for the brookite → rutile transformation of the monodispersed brookite particles may be expressed as $k = 1.28 \times 10^4 \exp(-\dfrac{17248 \pm 144}{T})$ s^{-1}.

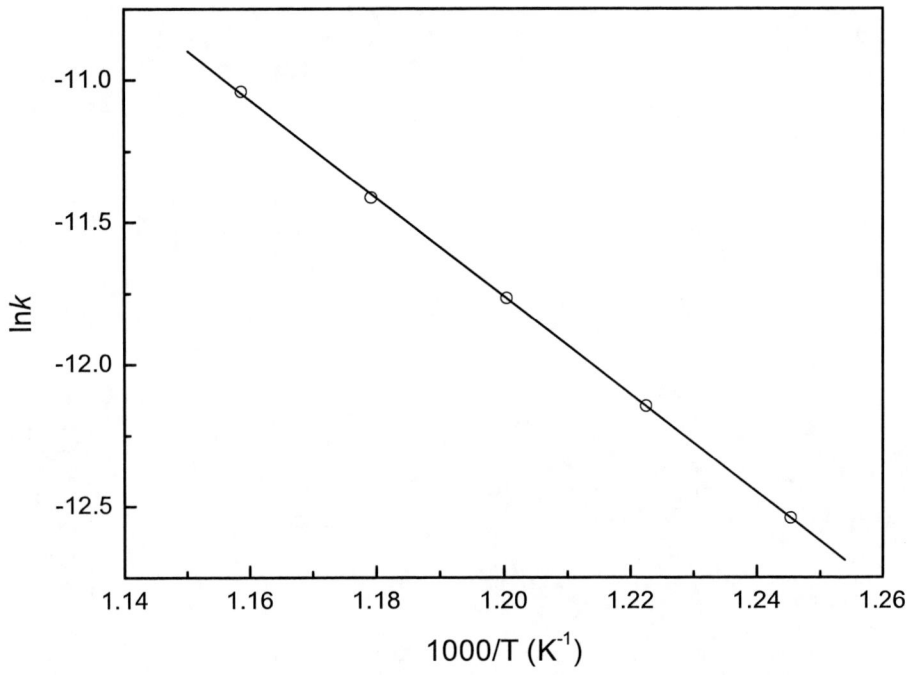

Figure 16. The rate factor (*k*) against reciprocal temperature on a ln-ln plot.

THE EFFECTS OF TITANIUM SOURCE AND PRECIPITANT ON MORPHOLOGIES OF ANATASE NANOCRYSTALS

It is interesting to note that replacing titanium trichloride ($TiCl_3$) with ammonium flurotitanate ((NH_4)$_2TiCl_6$, AFT) as the titanium source while keeping urea as the pH regulator yields quasi monodispersed spheres of anatase [20], instead of brookite (Figure 17). Here AFT and urea have their respective concentrations of 0.015 mol/L and 0.5 mol/L, identical to those used for synthesizing the brookite spheres shown in Figure 12a, and the reactions are also similarly performed under atmospheric pressure at ~83-100 °C. These anatase spheres are well crystallized despite of the low synthetic temperature, which is in contrast to the amorphous spheres obtained via

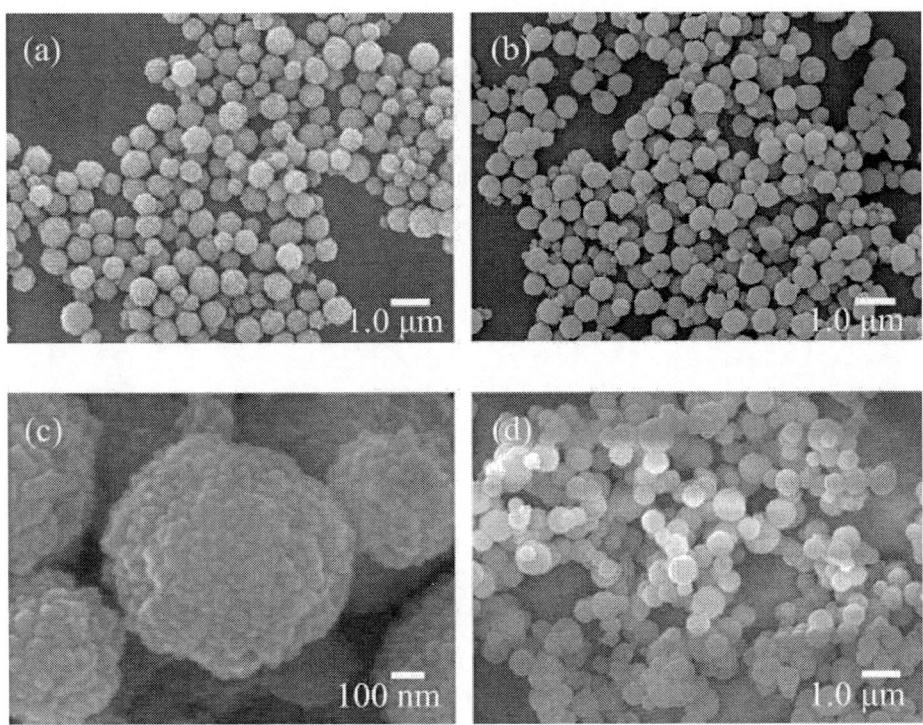

Figure 17. FE-SEM micrographs showing morphologies of the anatase spheres obtained by reacting mixed solutions of ammonium flurotitanate (0.015 mol/L) and urea (0.5 mol/L) for 30 min at (a) 83 °C, (b) 90 °C, and (d) 100 °C. (c) is the enlarged view of sample (b).

hydrolysis of titanium alkoxides. It is clear from Figure 17c that these spherical particles are built up of rounded nanoparticles of anatase, showing a structural feature similar to the brookite spheres shown in Figure 12a. Samples (a), (b) and (c) in Figure 17 have approximate particle sizes of 700-800, 600, and 500 nm, respectively, as assayed form the FE-SEM micrographs. These anatase submicron spheres are also formed via the aggregation mechanism discussed earlier. A higher reaction temperature tends to yield bigger primary crystallites, and samples (a), (b) and (c) have their respective crystallite sizes of ~15.2, 21.9 and 38.8 nm at 83, 90 and 100 °C. The crystallization of anatase rather than brookite is primarily owing to the higher pH (~7.0) of the current reaction system (as opposed to ~2.0 for brookite), though the existence of different supporting anions (F^- instead of Cl^-) may have also contributed to the different phase preference. Besides, the current reaction system does not involve the oxidation of Ti^{3+}, and thus the nucleation/growth would be significantly faster, favoring anatase crystallization. The anatase samples shown in Figure 17 are found to have bandgaps of ~3.23 eV via UV-vis spectroscopy, in close vicinity to that (~3.2 eV) widely observed for stoichiometric anatase crystals.

With ammonium flurotitanate as the titanium source, the effects of precipitant on morphologies of the resultant TiO_2 particles obtained via titration are investigated. Through dropwise addition of 400 mL of a 0.02 mol/L alkaline solution (ammonium hydroxide, sodium hydroxide, or tetramethyl ammonium hydroxide) to magnetically stirred 400 mL of a 0.005 mol/L ammonium fluotitanate solution at 90 °C, rice-shaped nanoparticles are resulted (Figure 18) irrespective of the precipitant used.

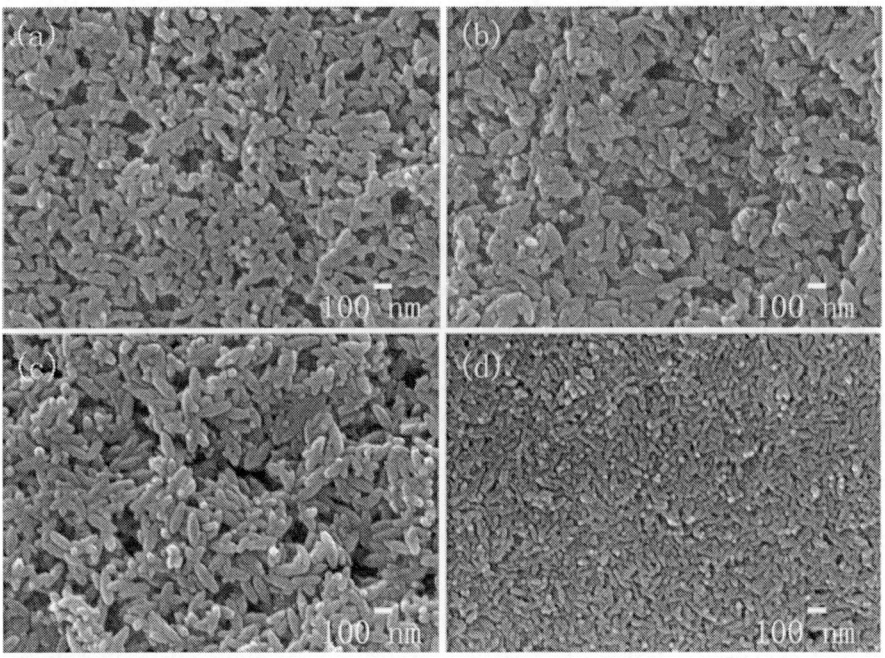

Figure 18. SEM images of the as-prepared TiO_2 particles synthesized with: (a) ammonium hydroxide, R=4:1, (b) sodium hydroxide, R=4:1, (c) tetramethyl ammonium hydroxide, R=4:1, (d) sodium hydroxide, R=8:1. R denotes the precipitant/ ammonium fluotitanate molar ratio.

Sizes of the nanoparticles are mainly in the range 100-150 nm when the precipitant/Ti^{4+} molar ratio (R) is kept constant at 4:1, and the size can be further reduced to below 100 nm by raising the R value to 8:1 (Figure 18d). X-ray analysis indicates that these rice-grain shaped particles are of pure anatase, without any trace of rutile or brookite impurity being detected (Figure 19). It is also notable that the nanoparticles are well dispersed and readily form dense films on glass substrates simply by evaporating their alcoholic/aqueous colloidal suspensions, allowing their potential applications in a wide range of technological fields.

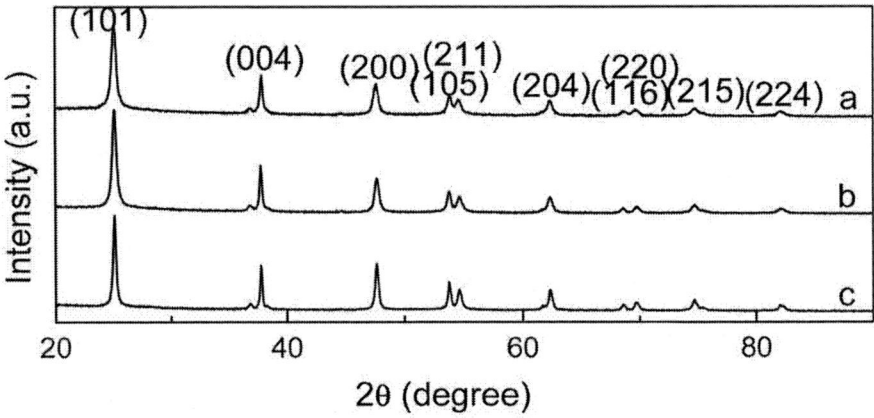

Figure 19. XRD patterns of the TiO_2 particles synthesized with: (a) ammonium hydroxide, (b) sodium hydroxide, and (c) tetramethyl ammonium hydroxide. The precipitant/ammonium flurotitanate molar ratio is 4:1 in each case.

The anatase nanoparticles made via titration (Figure 18) exhibit morphologies significantly different from those synthesized in the presence of urea (urea-based homogeneous precipitation, Figures 12a, 17), though they appear similarly uniform in size and shape. The changed morphologies are primarily due to the differed nucleation/growth rates among these synthetic techniques. Urea is known to undergo hydrolysis at elevated temperatures to give off carbon dioxide and ammonia, and this raises solution pH via ammonia dissolution in water. The hydrolysis of urea is rather slow at temperatures up to ~100 °C, which allows a low nucleation density. Bigger spheres are thus generated from the rather limited number of nuclei via gradual accumulation (aggregation) of the primary particles. Precipitation via titration, on the other hand, is characterized by rapid pH increase and extremely high nucleation density, and thus yields TiO_2 particles of significantly reduced sizes. The above results signify the importance of reactant selection in the phase structure and morphology controllable processing of TiO_2 nanomaterials.

Chapter 5

SINGLE MOLECULAR DESIGN OF NON-METAL DOPED TiO$_2$ FOR ENHANCED PHOTOCATALYSIS

While pristine TiO$_2$ nanocrystals have been finding broad technological applications, such as in photocatalysis, photovoltaic cells for solar energy harvesting, pigments, and ultrathin capacitors, enhanced and/or new functionalities may be derived via doping the material with either non-metallic or metallic elements.

Nanocrystalline TiO$_2$ remains one of the most promising photocatalysts up to date due to it higher efficiency, better stability, non-toxicity and availability. As a photocatalyst, one major disadvantage of TiO$_2$ is that it can only be activated by irradiation with UV light because of its relatively wide bandgap (~3.2 eV). As the UV light constitutes only ~5% of the solar energy compared to ~45% of the Vis light, any shift in the optical response from UV to the Vis spectral range will have a remarkable positive effect on the practical application of the material. There are two general approaches to achieve Vis responses of TiO$_2$: substituting the Ti sites of TiO$_2$ lattice with transition-metal cations (though frequently proved not effective) and doping TiO$_2$ with nonmetallic elements, typically nitrogen. The latter technique proves more promising as it may avoid deteriorating the thermal stability of the TiO$_2$ lattice and especially the possible increase in the amount of carrier-recombination centers. Following the pioneer work of Asahi et al. [21], a number of studies have confirmed that singly or multiply doping TiO_2 with non-metallic elements such as N, C, S, F, Cl, Br may extend the absorption edge into the Vis region and therefore enhances the catalytic performance of TiO_2 under Vis illumination in the degradation of harmful organic materials and in water splitting. Non-metallic doping of TiO_2 can be achieved by one of the following techniques: (1) sputtering followed by annealing under a controlled atmosphere, (2) direct oxidation of the dopant-containing titanium compounds (such as TiN, TiC, TiS_2) at a proper temperature, (3) annealing pure TiO_2 powders under a dopant-generating atmosphere (such as NH_3), and (4) solution-based strategies such as sol-gel and hydrothermal treatment. These techniques are straightforward but might have shortcomings of non-uniform dopant distribution in the TiO_2 lattice, as solid/gas reactions start at particle surfaces and therefore, due to kinetic reasons, the existence of gradient in dopant concentration is not easily avoidable.

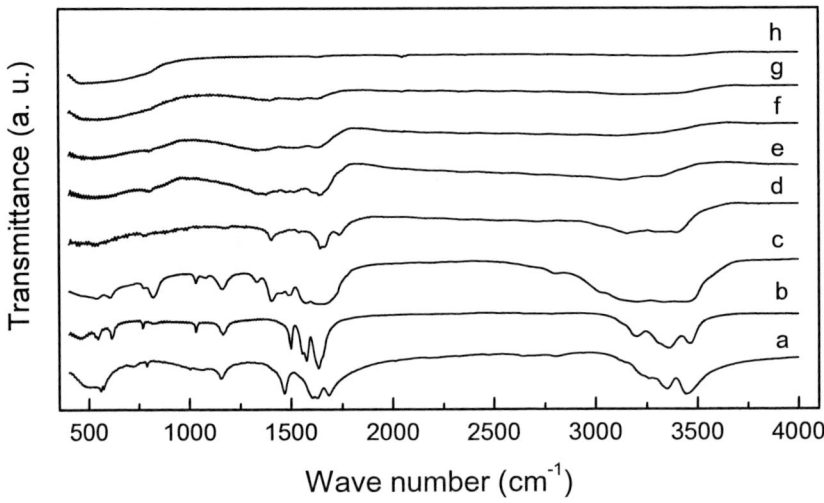

Figure 20. FTIR spectra of pure urea (a), the molecular precursor (b), and the products obtained by pyrolyzing the precursor for 2 h at 150 °C (c), 200 °C (d), 250 °C (e), 300 °C (f), 350 °C (g), and 400 °C (h).

Single molecular precursor, which provides an atomic mixing of titanium with the dopant elements, is considered ideal for an effective and uniform doping. In view of this, urea coordinated titanium trichloride, $Ti^{III}[OC(NH_2)_2]_6Cl_3$, is synthesized via reacting hot alcoholic solutions of urea and dry titanium trichloride ($TiCl_3$), and from which (C, N, Cl) codoped TiO_2 nanocrystals with high surface area (~98 m^2/g) is then obtained by pyrolyzing the precursor compound in air at ~450 °C [22]. Comparative studies show that the thus-made TiO_2 nanocrystals possess a substantially higher photocatalytic capability than Degussa P25 in the bleaching of methyl orange solution under Vis irradiation.

Table 1. Absorption maximums (cm^{-1}) for urea and the coordination complex [22]

Type of vibration	urea	$Ti^{III}[OC(NH_2)_2]_6Cl_3$
$v_{as}(NH_2)$	3435 (s)	3461 (s)
$v_s(NH_2)$	3346 (s)	3359 (s)
Third N-H band	3259 (sh)	3201 (s)
v (CO)	1684 (s)	/
v (CN)+δ(NH$_2$)	1628 (sh)	1632 (s)
v (CO)+ δ (NH$_2$)	1600 (s)	1574 (s)
v_{as}(CN)	1468 (s)	1499 (m)
ρ(NH$_2$)	1155 (m)	1163 (w)
v_s(CN)	1057 (vw)	1030 (w)
τ (ONCN)	787 (vw)	768 (vw)
δ (NCO)	/	614 (w)
δ (NCN)	557(s)	546 (w)

s, sh, m, w, vw denote strong, shoulder, medium strong, weak, and very weak, respectively.

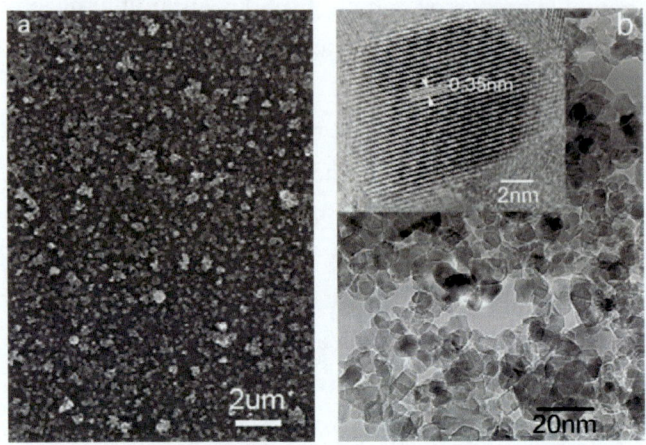

Figure 21. FE-SEM (a) and TEM (b) morphologies of the non-metallic doped TiO_2 nanocrystals pyrolyzed form the $Ti^{III}[OC(NH_2)_2]_6Cl_3$ coordination compound in air at 450 °C for 2 h. the inset in panel b shows the lattice fringe of a single anatase nanocrystallite.

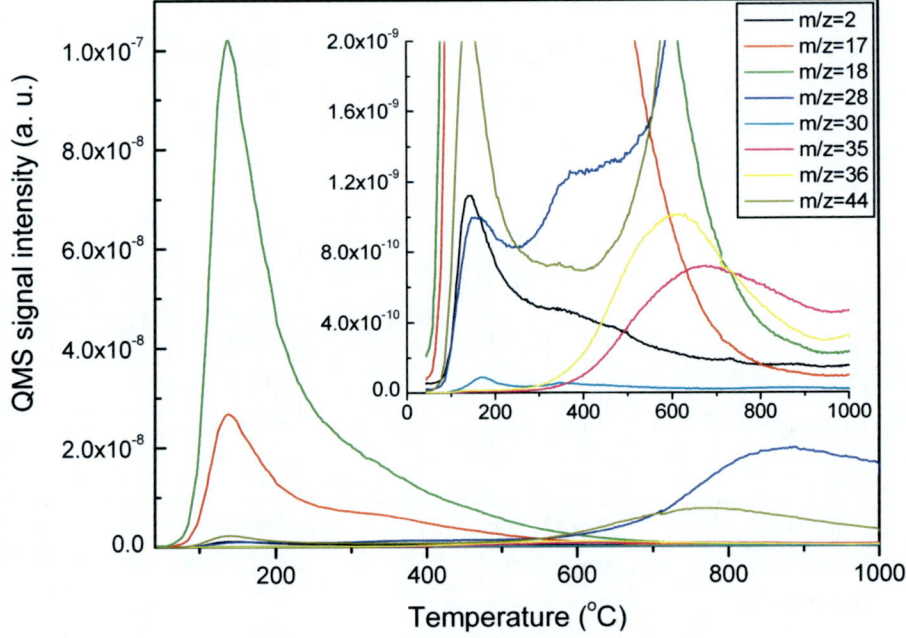

Figure 22. TDS spectra of the anatase TiO_2 powder pyrolyzed from the molecular precursor in still air at 450 °C for 2 h. Inset is the enlarged view of the spectra.

The as-synthesized coordination compound $Ti^{III}[OC(NH_2)_2]_6Cl_3$ has been well crystallized, but its XRD pattern could not match the data files of titanium compounds of known chemical compositions. Rietveld refinement [22] found that the coordination compound possesses a hexagonal crystal structure (space group: $P\bar{3}c1$) with $a=b=16.438(4)$ Å, $c=15.423(3)$ Å, $\alpha=\beta=90°$, $\gamma=120°$, and $V=3608.9(13)$ Å^3. Two possibilities of bond formation might be expected between the titanium (III) ions and urea molecules: oxygen (O)-to-Ti and nitrogen (N)-to-Ti. Urea-based complexes of other systems show that C=O→M (M: metal cations) coordination brings about only minor changes in many FTIR vibration bands of urea but frequently results in a shift of the 6 μm region to lower frequencies. The presence of N→M coordination, however, leads to a FTIR spectrum of the complex significantly different from that of free urea molecule. Figure 20 shows FTIR spectra of the coordination compound and urea, and assignments to the vibration bands are tabulated in Table 1. The formation of C=O→Ti bond is confirmed in this coordination compound, from the similarities of the FTIR spectra, the decreased wavenumber of the $\nu(CO) + \delta(NH_2)$ vibration and especially the absence of a carbonyl band at 1684 cm^{-1}. The FTIR spectra also indicate that annealing this precursor compound in air leads to successive thermal decomposition. At 400 °C only the Ti-O vibration is detectable, suggesting a complete conversion of the compound to TiO_2.

Figure 21 shows morphologies of the anatase powders obtained at 450 °C. FE-SEM observation (Figure 21a) indicates that the powders are composed of uniform soft agglomerates having sizes below 1 μm. TEM analysis (Figure 21b) reveals that the single crystallites are all sharp-edged, implying their excellent crystallinity as also evidenced by the well-resolved lattice fringes (the inset). The *d*-spacing of 0.35 nm of the inset single crystallite corresponds to the (101) plane of the anatase phase. The crystallite size averaged from randomly selected 100 crystallites in the TEM micrograph is 12.3 nm.

The existence of non-metallic dopants in the anatase lattice is confirmed by thermal desorption spectroscopy (TDS). TDS proves a useful tool in identifying origins of the desorbed species. For nanocrystalline TiO_2, our experiences revealed that physically adsorbed surface species, chemically bonded surface species, and lattice species show desorption maximums below ~200 °C, at ~400 °C, and at higher temperatures, respectively. Figure 22 shows TDS spectra of the anatase powder pyrolyzed at 450 °C. The signal peaking at ~137 °C with m/z=18 indicates the existence of surface adsorbed water. The m/z=17 signal has an intensity 23% of that of the m/z=18 signal, implying that it is also from water rather than from NH_3. Except for the case of surface water, desorption of gaseous species largely occurs at temperatures above 400 °C, in sharp contrast to the case of ammonia-treated P25 (Figure 23), confirming the existence of lattice dopants. Carbon, at least in part, is desorbed as CO_2 (m/z=44) and shows a desorption maximum at 780 °C. Lattice Cl is desorbed as HCl (m/z=36) and Cl (m/z=35), which show maximums at 610 °C and 675 °C, respectively (see the inset). The formation of HCl is due to the existence of desorbed H_2O in the investigated temperature range. The signal with m/z=28, peaking at ~885 °C, may come from N_2 or CO or a combination of the two, which is not separable via TDS. Nonetheless, lattice N, if there is any, is hardly desorbed as NO (m/z=30). It can also be inferred from the TDS spectra that Cl is less stable than C or N in the TiO_2 lattice. To derive the content of the lattice dopant, the powders pyrolyzed at 450 °C and 500 °C have been heated under TDS conditions up to 400 °C and then held there for 30 min to evaporate the surface adsorbed species. Both the powders are kept under vacuum before elemental analysis. Chemical analysis found 0.31 wt% of C, 0.59 wt% of N, and 0.11 wt% of Cl for the 450 °C treated powder, while 0.23 wt% of C, 0.52 wt% of N, and 0.08 wt% of Cl for the 500 °C treated powder. Based upon these analysis results, the powders pyrolyzed at 450 °C and 500 °C may roughly be expressed as $TiO_{1.9433}C_{0.0206}N_{0.0336}Cl_{0.0025}$ (total lattice dopants: 5.67at%) and $TiO_{1.9532}C_{0.01532}N_{0.0297}Cl_{0.0018}$ (total lattice dopants: 4.68at%), respectively.

As for the Degussa P25 powder treated with ammonia at 500 °C for 2h, TDS shows that the physically adsorbed NH_3 and chemically bonded surface nitrogen comprises a major portion of the total nitrogen content (Figure 23). These differences suggest that the single molecular route is significantly more efficient for the non-metallic doping of TiO_2 lattice.

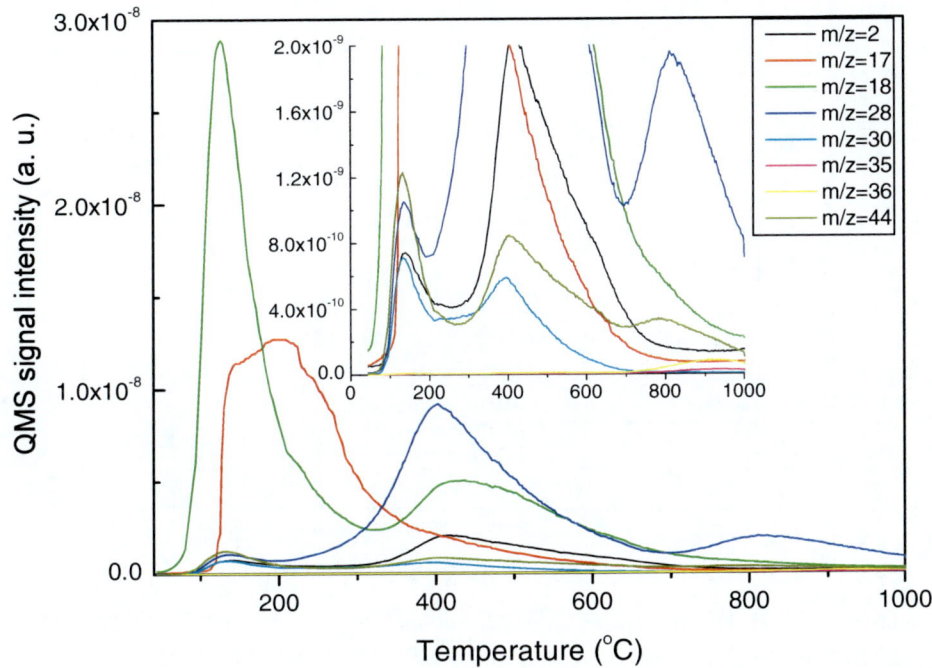

Figure 23. TDS spectra of the Degussa P25 powder treated under an ammonia stream (300 ml/min) at 500 °C for 2 h. Notice the different desorption maximums for the physically absorbed surface species (below ~200 °C), chemically bonded surface species (at ~400 °C), and lattice species (higher temperatures). The heating rate for the measurement is 0.5 °C/sec. Desorption behaviors of the m/z=17, 28, 35, 36, and 44 species of this sample are clearly different from those of the anatase powder pyrolyzed from the molecular precursor at 450 °C (Figure 22).

Figure 24 shows UV-vis absorption spectra of the TiO_2 powders derived from the molecular precursor, with that of P25 included for comparison. The P25 powder, appearing pure white, shows an onset of absorption at ~405 nm as commonly observed. Clearly, the incorporation of C, N, and Cl in the anatase lattice greatly enhances the light absorption in the UV-vis electromagnetic spectrum and effectively extends the absorption edge into the Vis range. The 450 °C powder shows slightly higher absorption than the 500 °C powder, owing to its higher dopant content. The 600 °C powder behaves quite differently, due to its partial transformation (76.4 wt%) to rutile and the greatly decreased dopant concentration at this higher temperature.

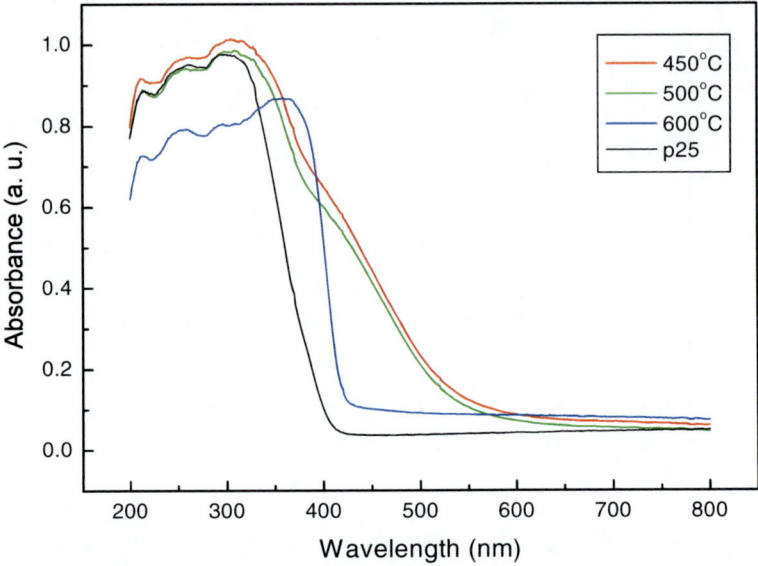

Figure 24. UV-Vis absorption spectra of P25 and the TiO_2 powders pyrolyzed from the molecular precursor in still air. The annealing time is 2 h at each temperature.

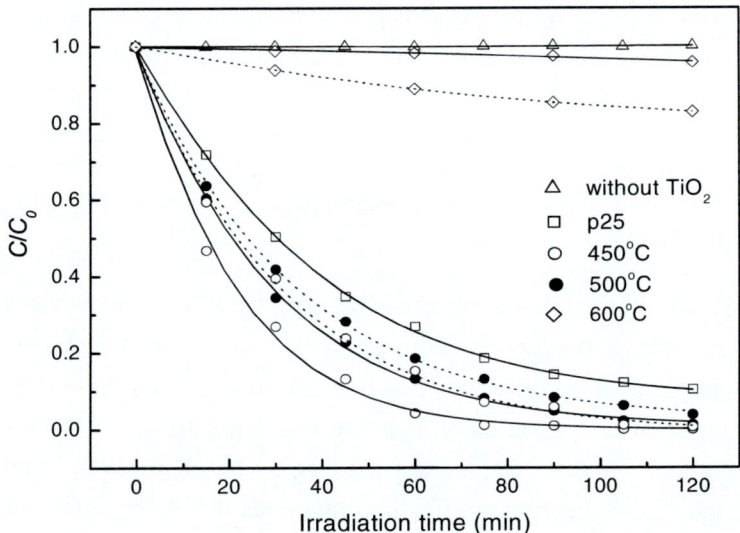

Figure 25. Photocatalytic performances of P25 and the TiO_2 powders pyrolyzed from the molecular precursor, evaluated via bleaching 20 μM methyl orange solutions under Vis irradiation. Solid lines: sample weight kept at 20 mg; dotted line: surface area of the loaded TiO_2 set as that of 20 mg of P25 (0.974 m^2/g).

Photocatalytic capability of the TiO_2 powder pyrolyzed from the $Ti^{III}[OC(NH_2)_2]_6Cl_3$ coordination compound is compared with that of P25 via bleaching 10 mL of 20 μM methyl orange solution. The P25 reference used in this work has a specific surface area of 48.7 m^2/g and is composed of ~83 wt% of anatase and 17 wt% of rutile, whose crystallite sizes are assayed from the Scherrer equation to be ~28 nm and 42 nm, respectively. P25 is one of the most powerful commercially available photocatalyst, and its excellent activity is proposed to be due to the synergetic contributions by mixing of anatase and rutile and the suppressing of hole/electron recombination by the presence of trace Fe^{3+} dopant. The use of P25 as a reference is hence of high benchmark. Two kinds of tests are performed in this work: fixed weight and fixed surface area of the loaded TiO_2 samples. A blank test indicates that in the absence of TiO_2 photocatalyst decoloration of the dye solution is negligible within the tested period of 2 h. In the presence of TiO_2 nanocrystallites, irradiation of the suspension with Vis light results in decoloration in all the cases, suggesting destruction of the absorption band of methyl orange. The 450 °C and 500 °C anatase powders show substantially higher efficiencies than P25 when the sample weight is fixed at 20 mg (Figure 25, solid lines). A complete decoloration of the dye solution is achievable within 2 h with both the anatase powders pyrolyzed from the molecular precursor but not with P25. The 450 °C powder shows the highest efficiency, causing a complete decoloration of the dye solution within 75 min. Specific surface area may play an important role in photocatalysis, as the reactions take place at particle surfaces. To exclude the effects of surface area, tests have also been made by setting surface areas of the loaded TiO_2 samples as that (0.974 m^2/g) of 20 mg of P25 (Figure 25, dotted lines). Apparently the 450 °C and 500 °C anatase powders are still more efficient than P25, implying that the higher efficiencies of these two powders mainly arise from their lowered bandgaps, which allows a more efficient absorption of the Vis light. The 450 °C and 500 °C powders show narrowed differences in their efficiencies at the same surface areas of the loaded samples, conforming to their similar bandgaps (Figure 24). The much lower efficiency of the 600 °C powder may mainly arise from its coarsened crystallite size, greatly decreased dopant concentration and especially its high content (76.4 wt%) of the rutile phase. Rutile is widely known to be less reactive than anatase in photocatalytic reactions.

Chapter 6

EFFICIENT DOPING OF TiO$_2$ NANOCRYSTALS VIA RADIO-FREQUENCY (RF) THERMAL PLASMA PROCESSING

Radio-frequency (RF) thermal plasma processing has similarities to flame pyrolysis but is characterized by its much higher processing temperature (~10^4 K) and its much faster quenching (10^{5-6} K/s) rate at the tail part of the plasma plume [23]. The technology has been finding wide applications in spheroidization, purification and alloying of refractory metallic or inorganic materials and also shows increasing potential in synthesizing nanoparticles of wide chemical compositions. Its high processing temperature makes it possible to use starting materials of various forms (solid, liquid, and gas) [16,24,25]. The fast quenching, on the other hand, makes thermal plasma a unique reaction field for efficient doping of the parent material with a variety of elements for a wider range of functionalities. Besides, phase structure and particle size control of the resultant nano powders might be achieved to some extent through injecting additional gases to influence the temperature profile of the plasma plume, the trajectories of the particles in the plasma reactor, and the cooling rate at the tail part of the plasma plume [26].

CHLORINE DOPING FOR IMPROVED PHOTOCATALYTIC PERFORMANCE

Through injecting aqueous $TiCl_3$ solution into Ar/O_2 thermal plasma, Cl-doped TiO_2 nanocrystals of excellent photocatalytic performance can be generated in one single step [27]. The plasma apparatus for powder synthesis mainly consists of a water-cooled induction plasma torch (Model PL-50, TEKNA Plasma System Inc., Sherbrooke, QC, Canada), a 2-MHz radio frequency power supply system (Nihon Koshuha Co. Ltd., Yokohama, Japan), a water-cooled stainless steel reactor, and a stainless steel filter connecting the reactor and a vacuum pump for reactor-pressure control. Details of the experimental setup are demonstrated in Figure 26.

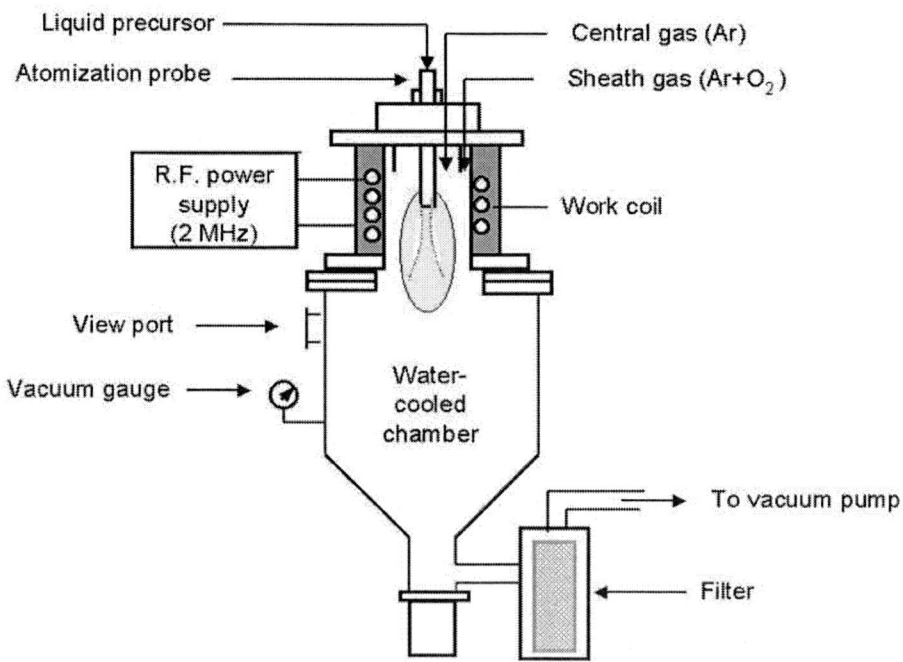

Figure 26. A schematic illustration of the experimental setup for thermal plasma processing.

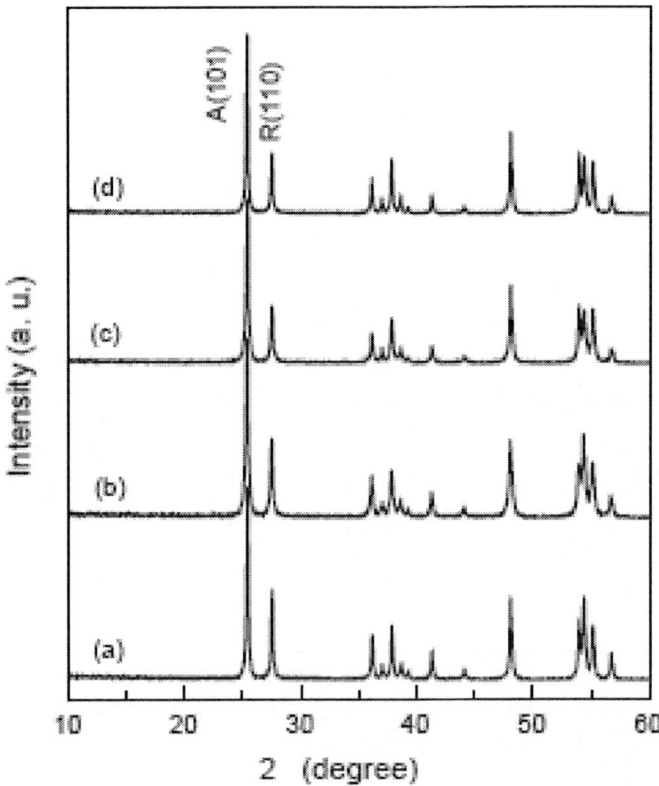

Figure 27. XRD patterns of the TiO$_2$ powders made via Ar/O$_2$ RF thermal plasma oxidation of TiCl$_3$ solution mists. (a)-(d) correspond to O$_2$ flow rates of 5, 25, 50, and 90 L/min, respectively.

The fundamental conditions for plasma generation are as follows: central gas, 30 L/min of Ar; sheath gas, 90 L/min of Ar/O$_2$ mixtures; atomization gas, 5 L/min of Ar; plate power, 25 kW; reactor pressure, 40 kPa. The TiCl$_3$ solution is delivered at 5.0 g/min by a peristaltic pump into the center of the plasma plume through an atomization probe (Model SA792-260-100, TEKNA Plasma System Inc.). The TiCl$_3$ solution is atomized into mists at the tip of the probe by the exiting Ar atomization gas (5 L/min) flowing through the probe. The O$_2$ gas in the plasma sheath serves as an oxidant for TiCl$_3$, and its flow rate is varied in the range 5-90 L/min to study its effects on powder properties.

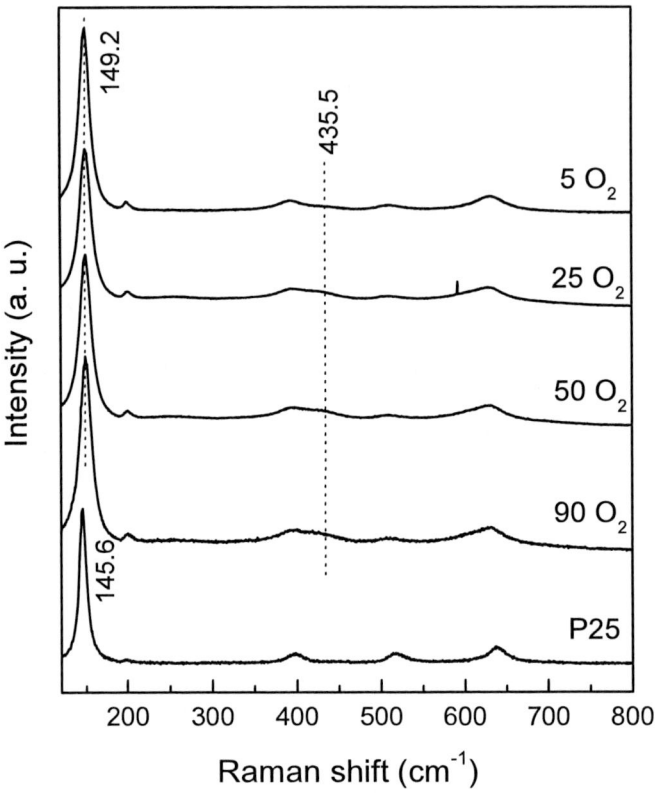

Figure 28. Raman spectra of the plasma generated TiO$_2$ nanopowders, with the O$_2$ flow rate (L/min) indicated.

The direct products of this plasma-assisted oxidation technique are well crystallized, thus saving the post-annealing step frequently needed in soft-chemical processing of TiO$_2$. No other phase except TiO$_2$ is identified via XRD (Figure 27), but the powders, irrespective of the O$_2$ flow rate in the plasma sheath (5-90 L/min), are mixtures of the anatase and rutile polymorphs. The anatase polymorph crystallizes as the major phase (~66-73 wt%) and neither the phase constituent nor the crystallite size exhibits a clear dependence on the O$_2$ flow rate. The main process of TiO$_2$ particle formation might be envisaged as follows. The TiCl$_3$ in the solution mists is rapidly oxidized by the Ar/O$_2$ thermal plasma, and the oxidation products undergo simultaneous evaporation due to the high plasma temperature (~10^4 K).

Depending on the trajectory and hence the thermal history in the plasma temperature field, the TiO_2 gas cluster may directly yield TiO_2 nanoparticles (gas-solid process) upon cooling and may produce melt droplets (melting point of TiO_2: 1870 °C). In this latter case, the TiO_2 nanoparticles may then be formed through nucleation/growth upon further cooling via a gas-liquid-solid process. The metastable anatase is widely observed to transform to thermodynamically stable rutile via conventional annealing at temperatures up to ~1000 °C. The formation of anatase rather than rutile as the major phase, despite the high processing temperature of the Ar/O_2 thermal plasma, is ascribed to the fast quenching at the plasma tail. Theoretical calculations of the free energy for nucleation suggest that a deeper cooling favors the nucleation of the metastable anatase phase [28,29]. Specific surface areas of the TiO_2 powders synthesized in this work fall in the range ~22-32 m^2/g, roughly half of that of the P25 powder (~50 m^2/g) made via flame pyrolysis of $TiCl_4$ vapors.

Figure 28 shows Raman spectra of the plasma-generated TiO_2 powders, with that of P25 included for comparison. In accordance with the results of XRD (Figure 27), these powders are apparently mixtures of the anatase and rutile polymorphs. The P25 powder exhibits its E_{Ag} at 145.6 cm^{-1} and E_{Rg} at 447.3 cm^{-1}, suggesting an almost stoichiometric composition. The plasma synthesized powders show their E_{Ag} and E_{Rg} at 149.2 cm^{-1} and 435.5 cm^{-1}, respectively, both shifted considerably from those of the stoichiometric TiO_2, indicating the existence of oxygen vacancies in both the anatase and rutile lattices. The extent of oxygen deficiency, however, is not appreciably affected by the O_2 flow rate in the plasma sheath used for powder synthesis, as these two characteristic Raman bands show fixed positions among the samples and do not vary from powder to powder. According to the calibration curves of the Raman spectrum of nanophase TiO_2 to the material's oxygen stoichiometry presented by Parker et al. [4], the anatase and rutile crystallites in the plasma-generated TiO_2 nanopowders would both have an approximate chemical composition of $TiO_{1.98}$. Such oxygen vacancies arise from the chlorine that has been incorporated into the TiO_2 lattice, as shown by the following results of TDS analysis.

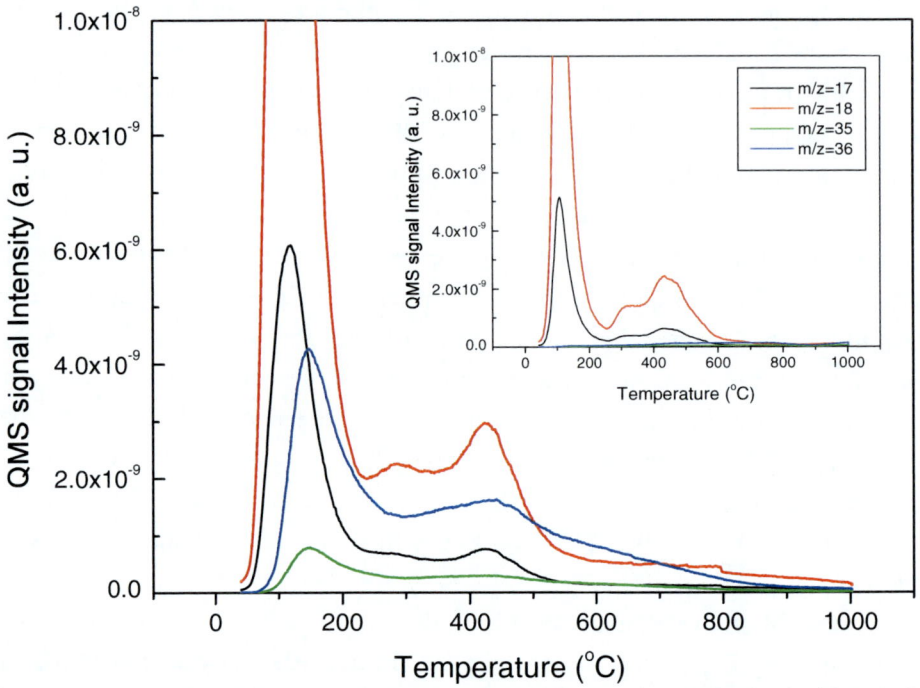

Figure 29. Thermal desorption spectra (TDS) of the TiO_2 powder synthesized with pure O_2 (90 L/min) as the plasma sheath. The inset is the TDS spectra of the commercial P25 powder.

Figure 29 shows TDS spectra of the TiO_2 powder synthesized with pure O_2 as the plasma sheath, with those of P25 included (the inset) for comparison. For both the powders, the signals centered at ~120 °C with m/z=18 arise from physically absorbed surface water. Again, the m/z=17 is also from H_2O rather than from NH_3. Chlorine is desorbed as HCl (m/z=36) and Cl (m/z=35), and the desorption occurs in two distinct temperature ranges. The desorption peaks centered around 150 °C are due to the presence of physically bonded surface chlorine, while those at higher temperatures of >300 °C from chemically bonded ones. It is also clear from the TDS spectra that the plasma-generated TiO_2 powder has significantly higher contents of physically and chemically bonded chlorine than P25 does. Semi-quantitative analysis of the desorption spectra in the range 300-1000 °C for chemically bonded chlorine yielded a value on the order of 10^{-2} (mole Cl per mole TiO_2) for the plasma synthesized TiO_2 powders, which is one order of magnitude higher than that (10^{-3}) for P25. The substantially higher chlorine contents are attributable to the much higher processing temperature of the thermal plasma than flame pyrolysis, which generates highly chemically reactive species and thus enhances chlorine doping.

TEM observations show that the plasma-generated TiO_2 particles typically have a broad size distribution. Figure 30 shows for example TEM morphology of the TiO_2 powder synthesized with pure O_2 (90 L/min) as the plasma sheath. Obviously, the majority of the particles have sizes under 100 nm but there is a small portion with diameters well above 1 μm. These two types of differently sized particles can be separated from each other via free sedimentation for 24 h of a suspension obtained by ultrasonically dispersing the total powder for 10 min in ethanol (solid loading: 1 vol%).

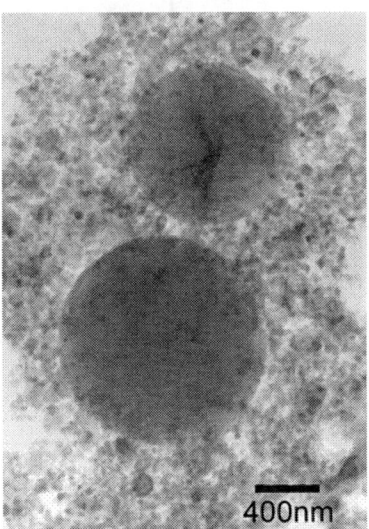

Figure 30. TEM micrograph showing a typical morphology of the plasma generated TiO_2 nanopowder. Notice the differently sized particles.

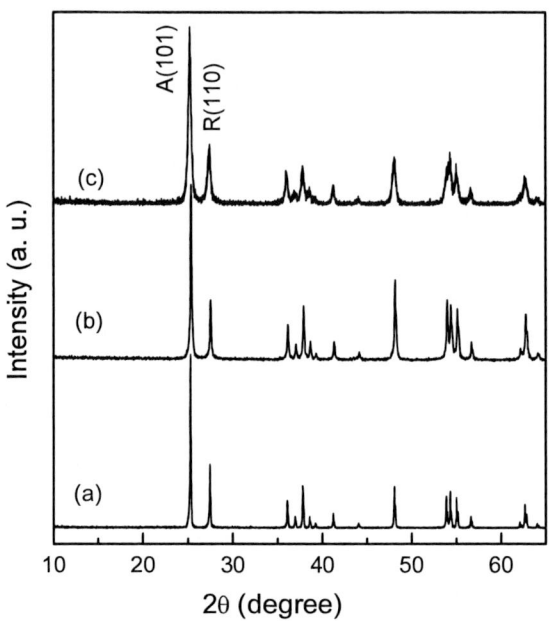

Figure 31. A comparison of the XRD patterns of the original (b) and the fractionated coarse (a) and fine (c) powders.

Table 2. Properties of the total and the fractionated powders

Sample ID	fraction (wt%)	rutile content (wt%)	anatase size (nm)	rutile size (nm)
a	34.1	32.1	91	96
b	100	27.9	65.1	66.1
c	65.9	26.6	42.4	46.8

The sample ID in this table corresponds to the labels in Figure 31.

Figure 31 compares XRD patterns of the total powder (Figure 31b) and the fractionated parts. Compared with the total powder, the fractionated fine portion (Figure 31c) shows slightly broadened XRD peaks while the coarse portion (Figure 31a) shows narrowed and sharpened XRD peaks. This suggests different average crystallite sizes for the two portions. Some properties of the three powders are summarized in Table 2, where it can be seen that fine particles constitutes a major portion (~66 wt%) of the whole powder and that the coarse portion has a slightly higher rutile content. Raman spectroscopy indicates that these three powders do not differ appreciably in oxygen stoichiometry, as they all exhibit the E_{Ag} at 149.2 cm^{-1} and the E_{Rg} at 435.5 cm^{-1}.

Morphologies of the fractionated powders are shown in Figure 32, and excellent dispersion is observed for both the coarse and the fine TiO$_2$ particles.

From the TEM micrographs it can be seen that the coarse particles are nearly perfect spheres with sizes ranging from submicron to ~5 μm (Figure 32a). For the fractionated fine part, particles of greater than 100 nm are hardly found and the tiniest crystallites are ~5 nm (Figure 32b). Additionally, all the fine crystallites tend to be faceted. These two distinctly different morphologies may imply different formation pathways of the particles, that is, the large spheres (Figure 32a) might have been formed via the gas-liquid-solid mechanism while the fine crystallites (Figure 32b) via a gas-solid route. Lattice spacing analysis via HR-TEM confirmed the existence of isolated nanocrystallites of rutile and anatase in the fine powder (Figures 32c, d). The well resolved lattice fringes suggest excellent crystallinity of the nanocrystallites, while the frequently observed tetragonal shapes may correspond well to the tetragonal crystal structures of anatase and rutile. For the fractionated coarse powder, the average crystallite sizes assayed from the Scherrer equation for anatase (~91 nm) and rutile (~96 nm) are much finer than the particle diameters (Figure 32a), which indicates that the particles are multi-crystalline. Micro-Raman spectroscopy of the micron-sized individual spheres indeed reveals that they are composites of inter-grown anatase and rutile [27].

Optical properties of the plasma generated TiO$_2$ nanocrystallites are studied via UV-vis spectroscopy, and is compared with that of Degussa P25 (Figure 33). Clearly, the plasma generated powders exhibit enhanced absorption in the UV regime and significantly red-shifted absorption edges. Regardless of the O$_2$ flow rate in the plasma sheath, these TiO$_2$ powders synthesized from TiCl$_3$ have indirect bandgaps of ~2.65 eV, lower than the 2.87 eV of P25. In addition, the almost identical bandgap values for all the plasma made TiO$_2$ suggest similar contents of lattice chlorine dopants, which is in agreement with the similar oxygen deficiency levels revealed via Raman spectroscopy (Figure 28).

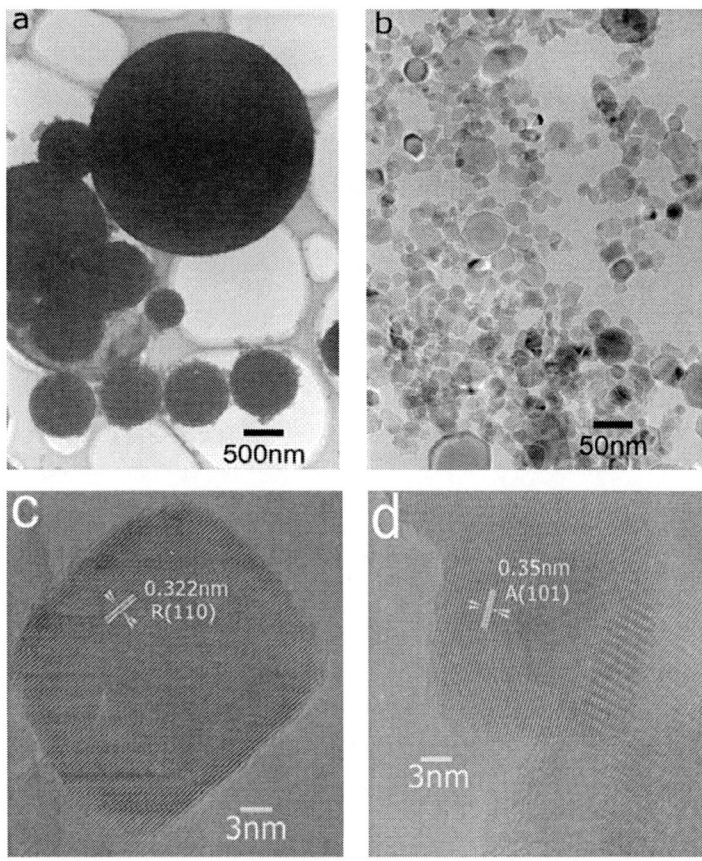

Figure 32. TEM micrographs showing morphologies of the fractionated coarse (a) and fine (b) TiO$_2$ particles. (c) and (d) are the HR-TEM lattice fringes of individual rutile and anatase nanocrystallites observed from the fine particles shown in part (b). A and R denote anatase and rutile, respectively.

A careful examination of the absorption spectra in the range 400-500 nm (Figure 33), within which wavelengths (405 and 436 nm) of the Vis light used for photocatalytic tests fall, shows that the absorption intensity increases steadily with decreasing O$_2$ flow rate in the plasma sheath during powder processing. As all the powders possess almost identical bandgaps, such a phenomenon is thus attributable to the different surface properties of the powders. This is plausible in view that the plasma atmosphere impacts particle surfaces more readily and more directly. At a lower O_2 partial pressure (lower O_2 flow rate in the plasma sheath), surfaces of the resultant TiO_2 particles may have more Ti^{3+} ions, leading to a higher absorption of the Vis light.

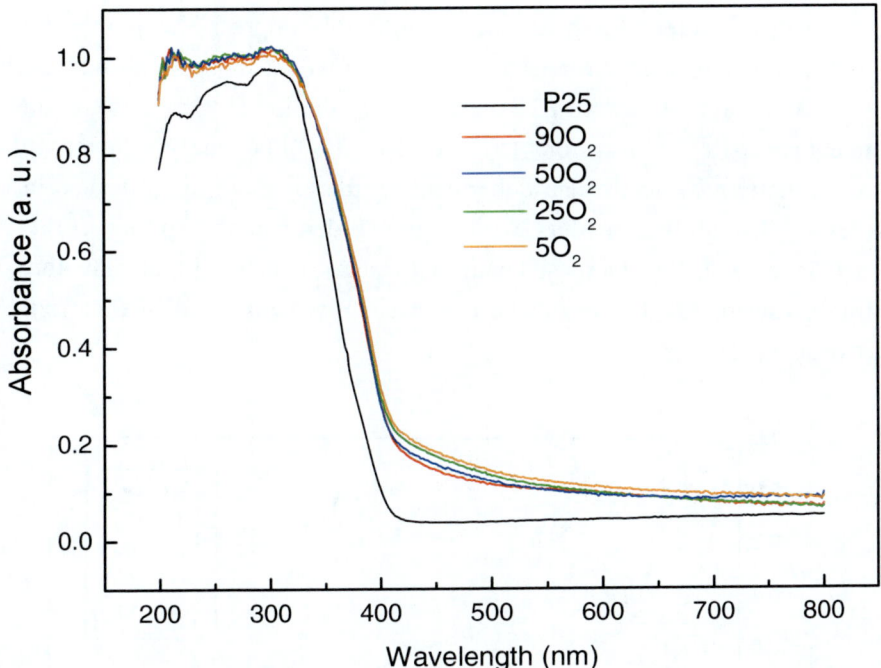

Figure 33. UV-vis absorption spectra for P25 and the plasma generated TiO_2 powders. The oxygen flow rate in the plasma sheath is indicated in the figure.

Photocatalytic performances of the chlorinated TiO_2 powders are compared with that of P25 via bleaching 20 µM methyl orange solutions under Vis (Figure 34a) and UV (Figure 34b) illumination. For the Vis tests, the surface area of all the tested powders has been set as that (0.5 m^2) of 10 mg of P25 by varying the sample weight to exclude the effects of specific surface area. For the UV tests, the TiO_2 samples have the same surface area as that of 1 mg of P25, and the intensity of the UV light used is 1 mW/cm^2 on the top surface of the suspension. From Figure 34a, appreciably higher efficiencies are confirmed for all the chlorinated powders made by thermal plasma, which is attributed to their narrowed bandgaps (Figure 33). Nonetheless, the performance of the powder shows a clear dependence on the O_2 input during powder synthesis, exhibiting a tendency opposite to the absorption intensity of the powder in the 400-500 nm range (Figure 33). This indicates that the excessive surface Ti^{3+} ions may have acted as surface recombination centers for h^+/e and accordingly suppress the photocatalytic activity. The plasma-synthesized TiO_2 powders also exhibit higher photocatalytic activities than P25 under UV irradiation (Figure 34b), owing to their enhanced UV absorptions (Figure 33). The photocatalytic reactivity of the chlorinated powder is also dependent on the O_2 input, as observed in the Vis tests. Since all the powders exhibit almost identical absorptions of the UV light (Figure 33), the decreased reactivity at a lower O_2 input may thus be similarly ascribed to the enhanced surface recombination of h^+/e carriers by the excessive Ti^{3+} ions.

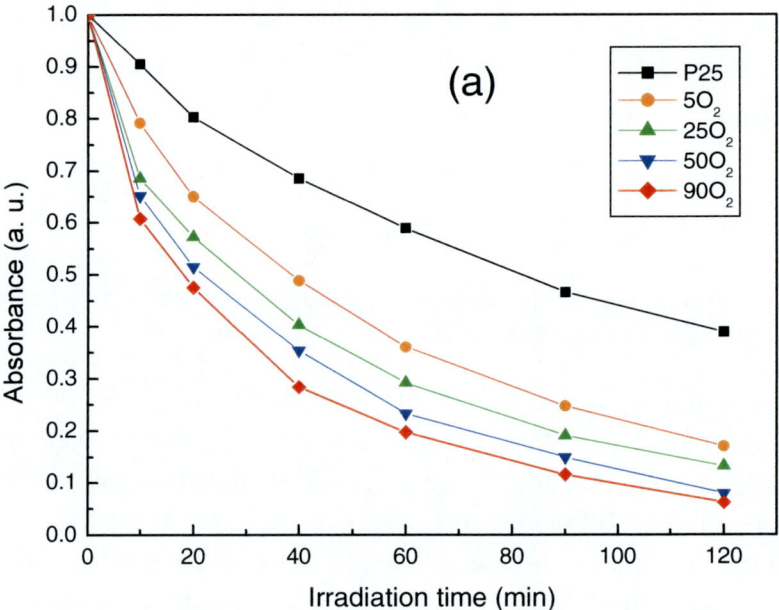

Figure 34. Continued on next page.

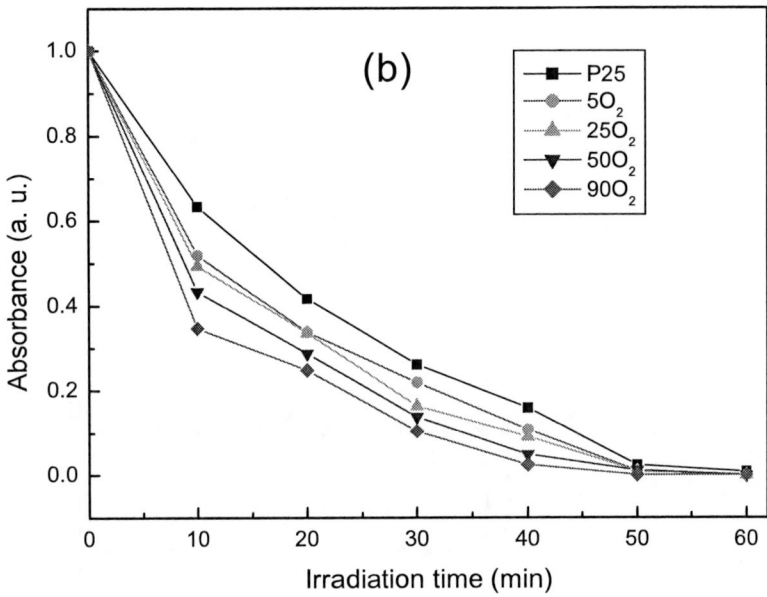

Figure 34. Degradation kinetics of 20 μM methyl orange solutions over TiO_2 photocatalysts under Vis (a) and UV (b) irradiation.

RARE-EARTH DOPING FOR NOVEL PHOTOLUMINESCENT PROPERTIES

With the thermal plasma processing technique, Eu^{3+}-doped TiO_2 nanocrystallites are synthesized by injecting an aqueous solution containing titanium tetra-*n*-butoxide ($Ti(OC_4H_9)_4$, TTBO) and europium nitrate ($Eu(NO_3)_3$) into the plasma plume [26]. TTBO undergoes immediate hydrolysis in the presence of water, and this is overcome by performing the following preparative procedures: (1) adding 0.1 mol (34 g) of TTBO to 0.4 mol (42 g) of diethanolamine ($HN(OC_2H_5)_2$, DEA, a chelate for Ti^{4+}) under magnetic stirring to obtain a clear solution (solution I). The chelating reaction between these two chemicals stabilizes the vivid reactive TTBO against hydrolysis, even in the presence of water; (2) dissolving europium (III) nitrate and citric acid ($C_6H_8O_7$, CA, a chelate for Eu^{3+}, CA:Eu^{3+}=1:1 in molar ratio) in

20 mL of distilled water to make a clear solution (solution II). Acids catalyze the hydrolysis of TTBO even in the presence of DEA and, to avoid this, the pH of solution II is adjusted to 9.0 with 3 mL of 25% ammonia solution; (3) mixing solutions I and II to yield the liquid precursor for plasma processing.

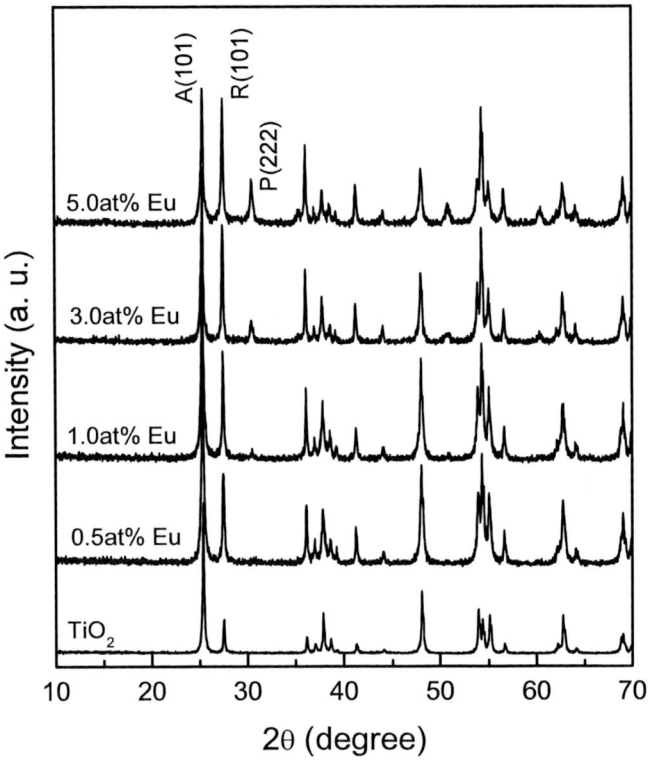

Figure 35. XRD patterns of the nanopowders synthesized under 40 L/min of O_2 input in the plasma sheath. The Eu^{3+} content of the precursor solution is indicated in the figure. A, R, and P denote anatase, rutile, and $Eu_2Ti_2O_7$ pyrochlore, respectively.

Elemental analysis on europium by standard ICP procedures confirmed that the prescribed Eu/(Ti+Eu) molar ratios in the precursor solutions have been kept to the final products. Figure 35 shows XRD patterns of the powders synthesized with 40 L/min of O_2 flow in the plasma sheath for some typical dopant concentrations. In all the cases the powders are mixtures of the anatase and rutile polymorphs as observed before. $Eu_2Ti_2O_7$ pyrochlore (JCPDS: No.

23-1072) appears in the powder doped with 0.5 at% of Eu^{3+}, and its diffraction intensity successively increases at a higher Eu^{3+} addition. This indicates that the solubility of Eu^{3+} in the TiO_2 lattice is rather limited, which is understandable from the rather large size mismatch between the Ti^{4+} (0.0605 nm for 6-fold coordination) and Eu^{3+} (0.0947 nm for 6-fold coordination) ions.

It is seen from Figure 35 that the addition of Eu^{3+} to the precursor solution has appreciable effects on the phase constituents of the products, which can be perceived from the relative intensities of the anatase (101) and rutile (110) peaks. The undoped powder has 22 wt% of rutile, which steadily increases to 52 wt% at 5 at% of Eu^{3+} addition. The average size of anatase crystallites keeps almost constant in the range 30-36 nm for all the powders, while the rutile size, varying between 64 and 83 nm in the studied range, is always much bigger than the anatase size. Changing the O_2 input in the plasma sheath does not affect significantly the phase constituent and crystallite size [28]. The enhanced rutile crystallization is primarily due to the creation of oxygen vacancies (for charge compensation) in the TiO_2 gas clusters upon Ti^{4+} is replaced with the subvalent Eu^{3+} ions [28,29].

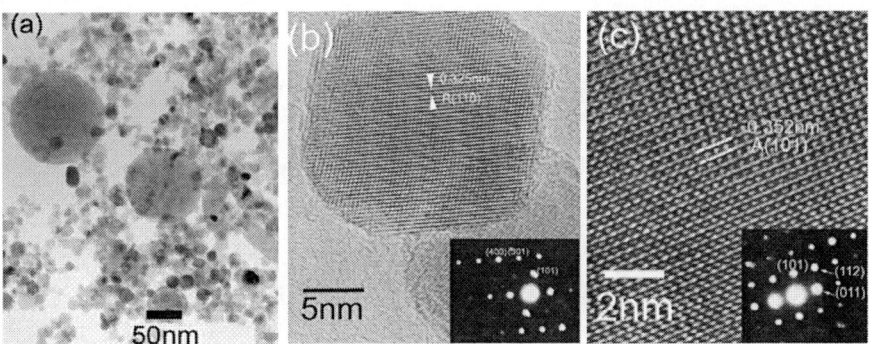

Figure 36. TEM micrographs showing: (a) overall morphology of the 0.5 at% Eu^{3+} doped TiO_2 particles synthesized with 40 L/min of O_2 input in the plasma sheath, (b) an individual rutile nanocrystallite, and (c) lattice fringe of an anatase crystallite. The insets in (b) and (c) are the SAED patterns of the nanocrystallites with zone axis of <0,-1,0> and <1,1,-1>, respectively.

Figure 36 exhibits overall morphology of the 0.5 at% Eu^{3+} doped TiO_2 nanopowder synthesized with 40 L/min of O_2 flow in the plasma sheath. The particles are dense and largely dispersed but a wide size distribution is observed, similar to those oxidized from TiCl$_3$ solution (Figure 30, 32). Separated crystallites of anatase and rutile can also be indentified through selected area diffraction and lattice fringe analysis.

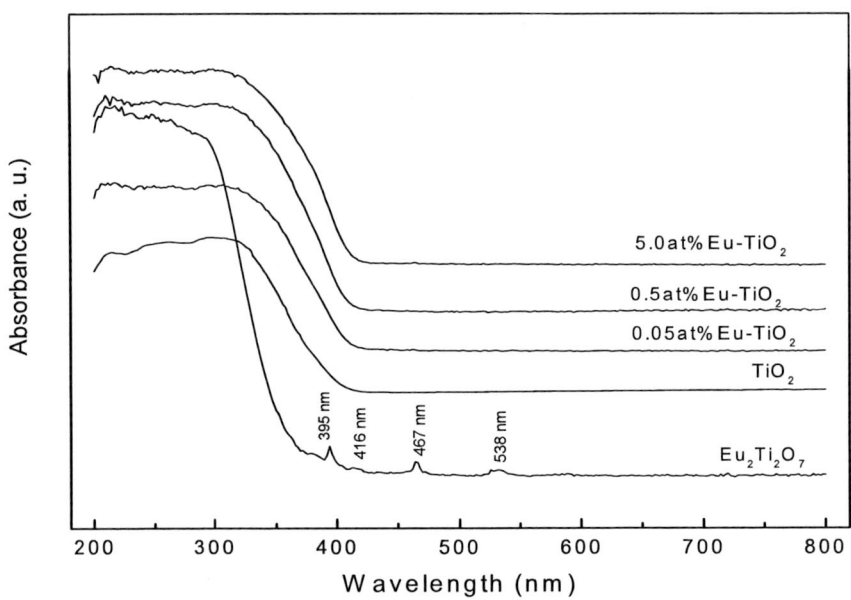

Figure 37. UV-Vis absorption spectra of the TiO$_2$:Eu^{3+} nanopowders, compared with those of TiO$_2$ and Eu$_2$Ti$_2$O$_7$ pyrochlore.

Figure 37 shows UV-Vis absorption spectra of the TiO$_2$:Eu^{3+} nanopowders, compared with the pure TiO$_2$ and Eu$_2$Ti$_2$O$_7$ synthesized analogously. Notice that data offsetting and profile enlargement (for Eu$_2$Ti$_2$O$_7$) have been made here for overall clearness of the figure. The pure TiO$_2$ sample exhibits an onset of absorption at 405 nm, corresponding to a bandgap of 3.06 eV. In addition to the absorption in the UV region, the pyrochlore powder shows additional absorptions at 395, 416, 467, and 538 nm, which are assignable to the intra-configurational 4f→4f transitions of Eu^{3+} ions. These absorption peaks should also appear in the UV-Vis spectra of the TiO$_2$:Eu^{3+} nanopowders, but are not clearly observed up to 5.0at% of Eu^{3+} due to the weakness of the peaks. Besides, the 395 nm peak occurs at a wavelength where the absorption by the TiO$_2$ host lattice is still relatively strong, and hence the absorption by Eu^{3+} ions at 395 nm might be quenched.

Figure 38(a) demonstrates excitation spectrum of the 0.5 at% Eu^{3+} doped TiO_2, measured by monitoring the 617-nm emission arising from the $^5D_0 \to {}^7F_2$ electronic transition of Eu^{3+} ions. The three peaks at 416, 467, and 538 nm, which show features similar to those observed in the UV-Vis spectrum of $Eu_2Ti_2O_7$, can be assigned to the $^7F_{0,1} \to {}^5D_3$, $^7F_{0,1} \to {}^5D_2$, and $^7F_{0,1} \to {}^5D_1$ transitions of the Eu^{3+} ions, respectively. The broad peak at 360 nm, whose position coincides well with the absorption band of pure TiO_2 (Figure 37), indicates that the Eu^{3+} ions can be effectively excited through the TiO_2 host lattice. Figures 38b-d show emission spectra of pure TiO_2 (Figure 38b) and the 0.5at% Eu^{3+} doped TiO_2 (Figures 38c, d). When excited at 467 nm, a wavelength longer than the absorption edge of TiO_2 (405 nm), the Eu^{3+}-doped nanopowder exhibits characteristic emissions in the range 550-750 nm (Figure 38c), which are associated with the electronic transitions from the excited $^5D_0$ level to the ^7F_j (j=1, 2, 3,4) levels of Eu^{3+} activators as commonly observed. Clearly, these emissions come from the absorption by the Eu^{3+} ions themselves, as confirmed by the excitation (Figure 38a) and the UV-Vis spectra (Figure 37). Upon UV excitation at 360 nm, a wavelength shorter than the absorption edge of TiO_2, the 0.5 at% Eu^{3+} doped sample shows luminescence from both the TiO_2 host and the Eu^{3+} ions (Figure 38d). It is noticed that the Eu^{3+} emission is clearly stronger than that from the host lattice and than that under 467 nm excitation. Furthermore, the emission from the TiO_2 host in Figure 38d is significantly weaker than that in Figure 38b. All these suggest an efficient nonradiative energy transfer from the TiO_2 host to the Eu^{3+} ions.

Figure 39 displays typical emission spectra of the TiO_2:Eu^{3+} nanoparticles as well as those of pure Eu_2O_3 and $Eu_2Ti_2O_7$ under 325-nm He-Cd laser excitation. It is the $^5D_0 \to {}^7F_2$ transition that gives a sharp red color when the laser beam impinged upon sample surfaces. The Eu^{3+}-doped samples exhibit emissions clearly different from those of Eu_2O_3 and $Eu_2Ti_2O_7$, in terms of peak positions and peak shapes, implying that the 617-nm red emission is indeed from the Eu^{3+} ions doped in the TiO_2 lattice instead of from isolated Eu_2O_3 or $Eu_2Ti_2O_7$. The intensity ratio of the $^5D_0 \to {}^7F_2$ transition (monitored at 617 nm) to the $^5D_0 \to {}^7F_1$ transition (monitored at 599 nm), called asymmetry factor, keeps almost constant at 9.7 for all the samples doped up to 0.5 at% of

Eu^{3+}, indicating that the overall Eu^{3+} local environments do not differ from sample to sample.

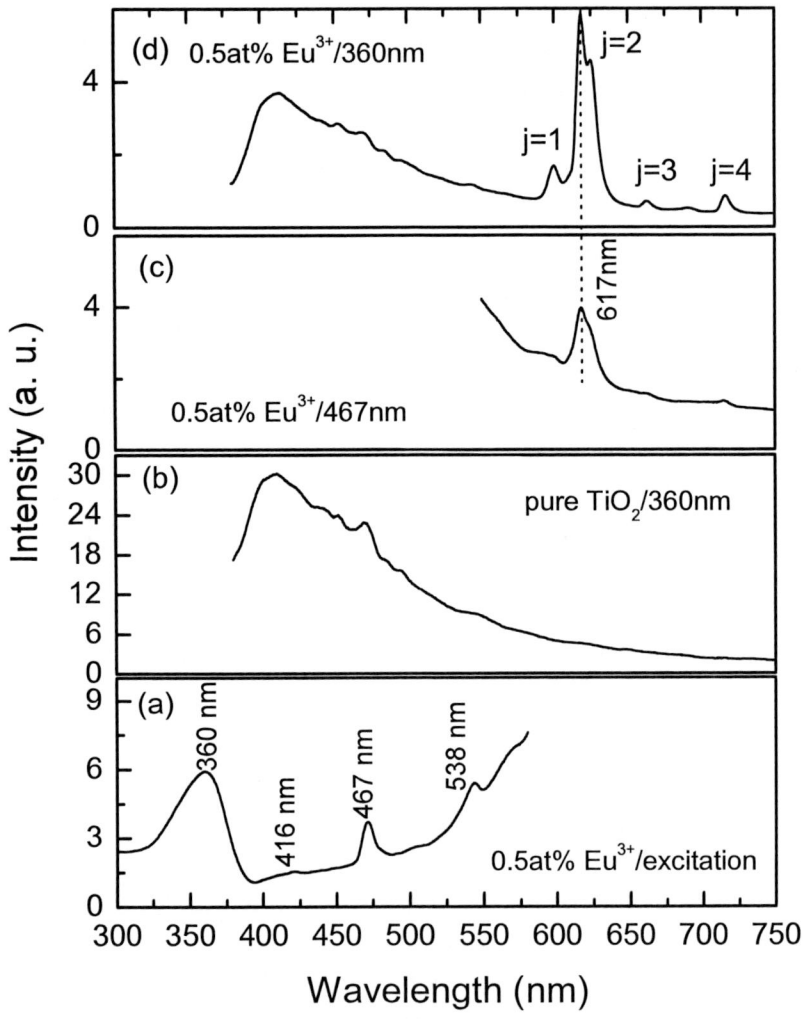

Figure 38. Excitation and emission spectra of the nanopowders. (a): excitation spectrum obtained by tracking the 617 nm emission of the 0.5 at% Eu^{3+} doped TiO$_2$; (b)-(d): emission spectra with sample composition and excitation wavelength indicated. Direct comparison of the emission intensities can be made among (b)-(d).

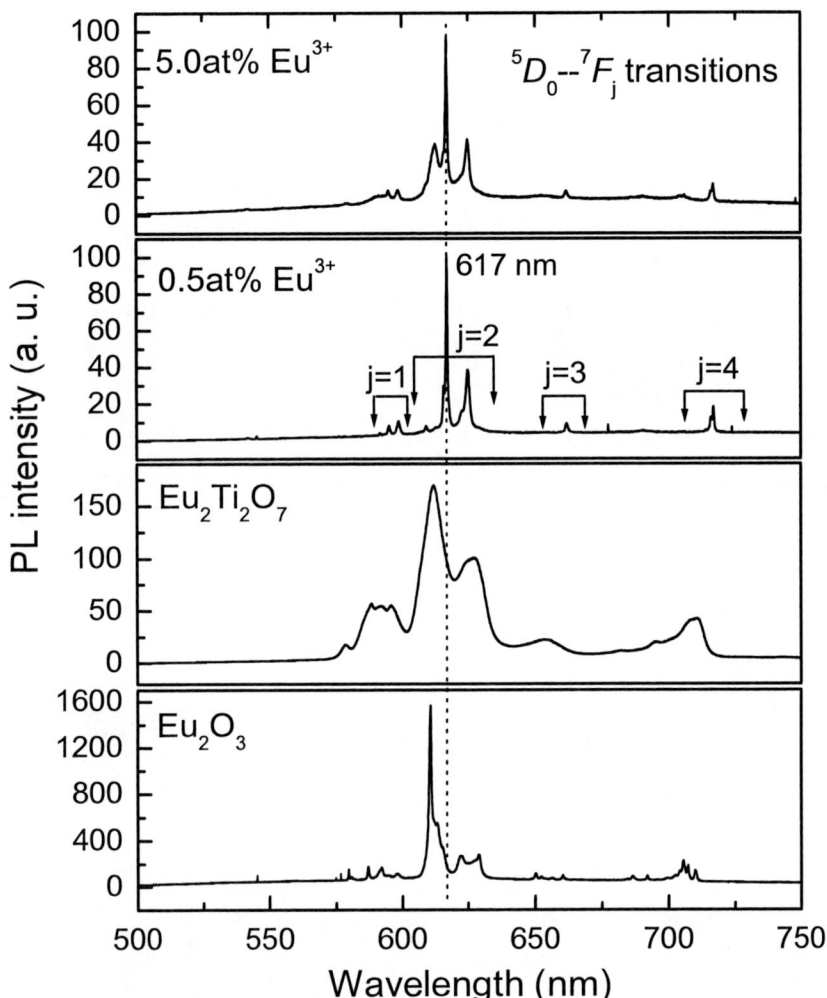

Figure 39. Emission spectra of the TiO$_2$:Eu^{3+} nanopowders under 325 nm He-Cd laser excitation, compared with those of Eu$_2$O$_3$ and Eu$_2$Ti$_2$O$_7$. For the four samples, the original PL signals have been divided by a common factor for clearer demonstration and easier intensity comparison. Satellite emissions (Stark splits) of each $^5D_0 \rightarrow {}^7F_j$ (j=1, 2, 3, 4) transition is indicated by an arrowed bracket in the figure.

Replacing Eu^{3+} with Er^{3+} as the activator, Er^{3+}-doped TiO$_2$ nanocrystallites that show the non-radiative energy transfer from the TiO$_2$ host lattice to the Er^{3+} luminescent centers are similarly produced in one step via RF thermal plasma processing [30]. Again, the solubility of Er^{3+} in the TiO_2 lattice is found to be around 0.5 at% and above which $Er_2Ti_2O_7$ pyrochlore is formed as an impurity phase. To show the advantages of this high-temperature RF thermal plasma processing technique, a sample with 0.25 at% of Er^{3+} is also made via "coprecipitation" for comparison. In this latter case, the precursor powder is precipitated by dripping at room temperature a concentrated ammonia solution (13 M) into a mixed solution of titanium trichloride ($TiCl_3$, 0.1 M) and erbium nitrate until a final pH of 10 is reached. After removing the byproducts by repeated washing with distilled water, the precursor is dried at 100 °C for 12 h and is then annealed in a tube furnace under flowing O_2 gas (100 mL/min) at 475 °C for 6 h. The annealed sample and the plasma generated ones have similar phase constituent (~80 wt% of anatase and 20 wt% of rutile) and similar crystallite size (~30 nm).

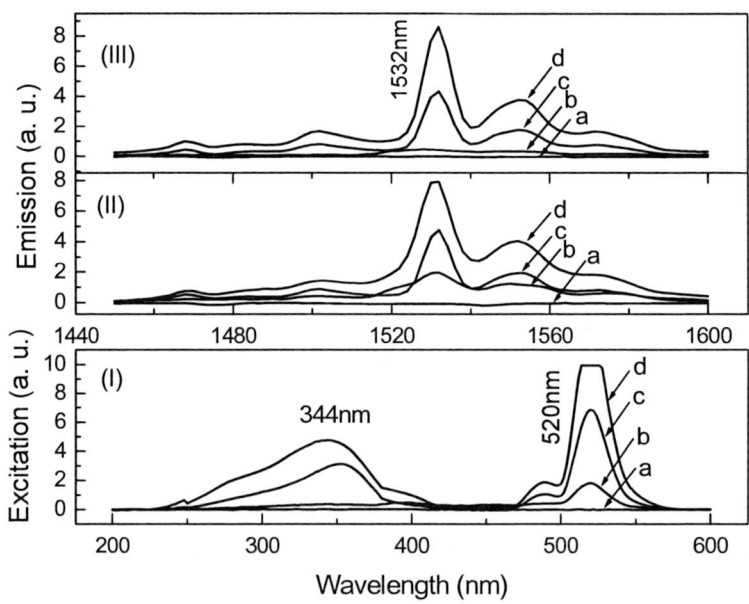

Figure 40. Excitation (I) and emission (II, III) spectra of the TiO_2:Er nanopowders. (I) was obtained by monitoring the 1532 nm emission, while (II) and (III) were obtained under 520-nm and 344-nm excitations, respectively. Samples a, c and d are synthesized by thermal plasma processing and with Er^{3+} contents of 0, 0.25, and 0.5 at%, respectively. Sample b is the 0.25 at% Er^{3+} doped nanopowder made via chemical coprecipitation.

Figure 40(II) shows PL spectra of the TiO$_2$:Er nanopowders under 520-nm excitation. The doped samples all exhibit the characteristic luminescence of trivalent erbium at ~1532 nm, while the undoped sample (line a) does not. Obviously, in this case the luminescence arises from the self-absorptions by Er^{3+} activators. It is noteworthy that, for the same Er^{3+} content of 0.25 at%, the plasma-generated nanopowder (line c) exhibits a luminescence intensity 2.1 times that of its "co-precipitated" counterpart (line b), as determined from the integrated area of the peaks in the range 1490-1600 nm. This shows advantages of the plasma processing technique. It is believed that surface hydroxyls, which partially remain even after annealing, are responsible for the lower luminescence intensity of the "co-precipitated" powder. For the plasma-synthesized nanopowders, the 0.5 at% sample (line d) has a luminescence intensity about 2 times that of the 0.25 at% one (line c). For both the powders, the 1532 nm emission has a full width at half maximum (FWHM) of 10 nm, which is smaller than those observed for the erbium-implanted SiO$_2$ (~13 nm), SR-350 resin (~21 nm), porous silicon (~23 nm), and Er-doped TiO$_2$ xerogels (15-30 nm) [30]. The sharper luminescence peaks may suggest better crystallinity of the plasma processed samples. Figure 40(I) shows excitation spectra of the samples obtained by monitoring the 1532 nm emission. Self-excitations are found at 520 nm for all the Er^{3+} doped samples, while pure TiO$_2$ (line a) exhibits no detectable excitation peak in the range 200-600 nm. For the same Er^{3+} concentration of 0.25 at%, the co-precipitated sample exhibits a much weaker excitation band at 520 nm (Figure 40(I), line b) than its plasma processed counterpart (Figure 40(I), line c). As measurements are performed under identical conditions, this phenomenon may possibly arise from (1) residual surface hydroxyls, which quench both PL excitation and emission, and (2) the different occurrence of Er^{3+} in the two samples. In the co-precipitated powder, The Er^{3+} ions may be enriched in a separate phase (though not detectable with XRD) instead of uniformly distributed in the TiO$_2$ lattice as dopants. Besides the 520-nm excitation, an additional peak centering at ~344 nm is observed for the plasma-generated nanopowders (lines c, d) but hardly for the "co-precipitated" one (line b). This provides direct evidence that energy transfer occurs from the TiO$_2$ host to Er^{3+} in the plasma-generated nanopowders while not in the co-precipitated and annealed one. Figure 40(III) shows PL spectra of the nanopowders under 344-nm excitation. Pure TiO_2 (line a) does not exhibit any luminescence at 1532 nm, which is also negligible for the sample made via co-precipitation (line b). The plasma-synthesized nanopowders (lines c, d) both exhibit sharp (FWHM=10 nm) luminescence at 1532 nm as expected from the energy transfer. Again, the luminescence intensity of the 0.5 at% sample is roughly twice that of the 0.25 at% one.

The non-radiative energy transfer in these rare-earth (RE^{3+}) doped TiO_2 nanocrystals may allow them to be used in optoelectronic devices, which need excitation of the activators with an electrical current. The energy transfer from the TiO_2 host to RE^{3+} ions is a defect mediated process, and a model of which has been proposed by Frindell et al. [31] for the RE^{3+}-doped mesoporous TiO_2 films. According to this model, UV light is absorbed in the bandgap of TiO_2 and the energy is relaxed to the defect states. As the defect energy levels of the TiO_2 host are higher than that of the emitting state of RE^{3+} ions, energy transfer to the crystal field states of RE^{3+} ions then takes places, resulting in an efficient photoluminescence from the plasma generated nanoparticles of TiO_2:Eu and TiO_2:Er. The defects commonly encountered in TiO_2 might be oxygen vacancies, interstitial or substitutional Ti^{3+} ions, interstitial Ti^{4+} ions, and so forth. In the present case, the energy mediating defects may dominantly be the oxygen vacancies generated by the substitutional doping of subvalent RE^{3+} ions into TiO_2 lattice.

TRANSITION METAL DOPING FOR ROOM TEMPERATURE FERROMAGNETISM

Since the discovery of room temperature ferromagnetism (RTF) in cobalt doped TiO_2 thin films by Matsumoto et al. [32], the physical properties of dilute magnetic semiconductors (DMS) have been drawing increased attentions due to their potential applications in optoelectronics, magnetoelectronics, spintronics, and microwave devices. Doping semiconducting oxides with transition metal (TM) ions in order to induce RTF has been one of the major challenges for many research groups, and to date a number of studies have reported on high-temperature ferromagnetism in oxide materials such as TM-doped TiO_2, ZnO, and SnO_2 (TM = Co, Ni, Cr, Mn, V, and Fe) [33]. For the cobalt-doped TiO_2 (TiO_2:Co) system, there has been a report showing that RTF can only be achieved in the anatase form while others showed RTF in rutile [34-36]. These contradictions suggest that the magnetic properties of TiO_2:Co heavily depend upon the methodologies and the conditions of sample preparation, as can be seen from Table 3, which summarizes some typical observations reported in recent years [33]. For DMS materials, one issue that still remains under debating is the origin of RTF. While a number of reports show its intrinsic nature, others claim that it might be extrinsic, that is, due to the existence of isolated metallic clusters of the dopant. Though both thin film and powdered forms have been prepared for ferromagnetic TiO_2:Co, it is noticed that in majority of the previous studies the samples are synthesized under vacuum or in reducing atmospheres, adding weight to the speculation that in these cases the observed RTF might be arising from metallic Co. In contrast to these previous studies, Co-doped TiO_2 nanocrystallites have been synthesized via Ar/O_2 RF thermal plasma oxidation of atomized liquid precursors containing titanium tetrabutoxide and cobalt (II) nitrate [33]. Such an oxidative atmosphere effectively prevents the possible formation of any metal elements, and thus gives a better understanding of the origin of the magnetic properties.

Hyper-powder XRD analysis (15 kW) shows that without Co-doping the TiO_2 nanopowder is a mixture of the anatase and rutile polymorphs, as repeatedly observed for the plasma generated TiO2 (Figure 41).

Phase constituents of the resultant TiO_2:Co nanopowders are given in Figure 42, where it can be seen that at 7 at% of Co^{2+} addition the powder is almost a pure rutile phase, with only trace amounts of anatase and $CoTiO_3$ detected. No peaks corresponding to metallic Co or binary cobalt oxides can be found in the XRD patterns even at a high Co/(Co+Ti) ratio of 0.50. Though difficult to determine unambiguously the solubility limit of cobalt in the TiO_2 lattice from these XRD patterns, there is no doubt to say that the excess cobalt forms $CoTiO_3$ and Co_2TiO_4 compounds, both are known to be non-ferromagnetic at room temperature. The Co_2TiO_4 phase (Co/Ti=2:1 molar ratio) is formed after CoTiO$_3$ (Co/Ti=1:1 molar ratio), which is in accordance with the gradually increased Co^{2+} content in the precursor solution. The formation of CoTiO$_3$ and Co$_2$TiO$_4$ titanates may also imply that cobalt most likely has an oxidation state of 2+ in the plasma generated TiO$_2$ nanopowders. The enhanced rutile crystallization by Co^{2+} doping is due to the increased amounts of oxygen vacancies in the TiO$_2$ lattice, which are created by replacing the Ti^{4+} sites of the TiO$_2$ lattice with subvalent Co^{2+} for charge compensation. Similar phenomena were previously observed upon doping TiO$_2$ with Fe^{3+} ions [29] or with RE^{3+} ions [28,30]. BET analysis found specific surface areas of 40-55 m^2/g for all the plasma generated powders, irrespective of the cobalt content, suggesting average particle sizes of 27-37 nm.

Table 3. Some typical recent reports on the TiO$_2$:Co system [33]

preparation technique	form	crystal structure	Co content (at%)	Ms (μ_B/Co)	remarks
MBE	film	anatase	7	0.32	as made
MBE	film	anatase	3	1.25	O$_2$-plasma assisted
MBE	film	rutile	5	1.0	as made
sputtering	film	rutile	4	0.94	as made
MBE	film	anatase	4	1.7	Co cluster found
MBE	film	anatase	7	1.55	Co cluster found
PLD	film	anatase	7	1.44	Co cluster@>2at% of Co
sol-gel	powder	anatase	10	4.1	after H$_2$-treatment@573K
SP	film	anatase	10	no RTF	ferromagnetic@≤5K
PLD	film	anatase	5	0.16	as made
SSR	powder	rutile	1	1.75	CoTiO$_3$ found@>2at% of Co
sputtering	film anatase	anatase	2	1.1	UHV annealed
Exfoliation	2D nanosheet	-	20	1.4	as made

MBE: molecular beam epitaxy; PLD: pulsed laser deposition; SP: spray pyrolysis; SSR: solid state reaction; UHV: ultrahigh vacuum.

Efficient Doping of TiO₂ Nanocrystals... 73

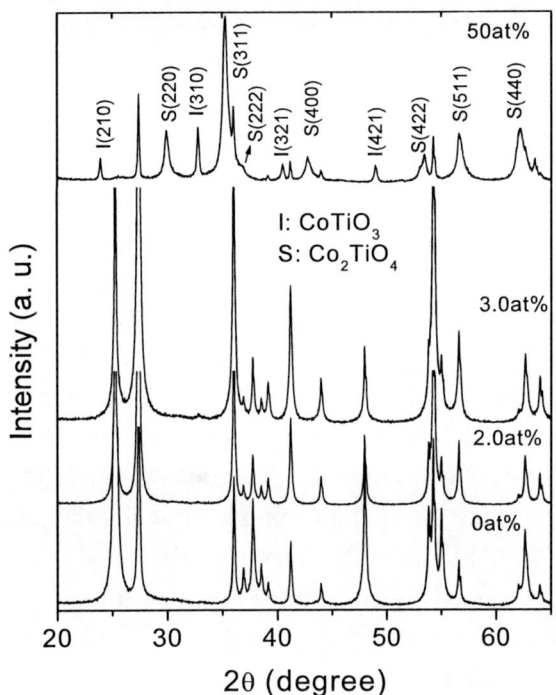

Figure 41. Hyper power XRD patterns of the plasma synthesized nanopowders, with cobalt contents indicated in the figure.

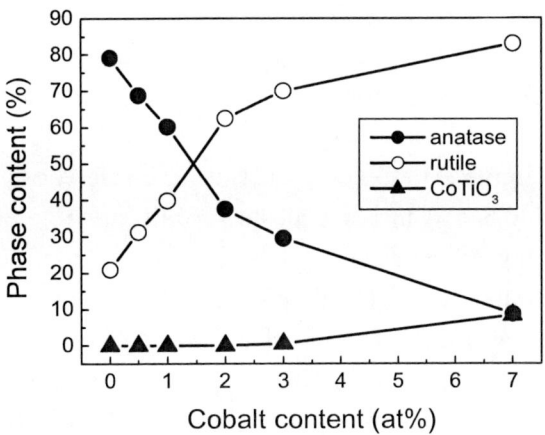

Figure 42. Phase constituents of the plasma generated nanopowders with up to 7 at% of Co^{2+} addition in the starting precursor solution.

Figure 43 shows Raman spectra of the powders with up to 7 at% of Co^{2+}-doping. The pure TiO_2 powder can be regarded as stoichiometric, judged from the anatase E_g mode at ~144 cm^{-1} and the rutile E_g mode at ~447 cm^{-1}. Raman band shifting is observed for both the anatase and rutile nanocrystals in the powders doped with up to ~2 at% of Co^{2+}. Above this doping level, however, band positions of both the TiO_2 polymorphs keep almost constant. Similar phenomena were previously observed in Fe^{3+}-doped TiO_2 [29], and these indicate that Co^{2+} largely substitutes for the Ti^{4+} sites of the TiO_2 lattice and that the solubility of Co^{2+} in TiO_2 is around 2 at%. The solubility limit determined here for Co^{2+} (~2 at%) is substantially smaller than that (~20 at%) for Fe^{3+} in TiO_2 [29], which might be understood from the following three aspects: (1) for 6-fold coordination, Co^{2+} (0.0745 nm, high spin state) is much bigger than Fe^{3+} (0.0645 nm, high spin state), while the later has an ionic size similar to Ti^{4+} (0.0605 nm); (2) every two Fe^{3+} ions while only one Co^{2+} is needed to create one oxygen vacancy for charge compensation, and hence within the tolerance of the TiO_2 lattice the amount of Co^{2+} that can be accommodated into TiO_2 is smaller; (3) cobalt oxide is more alkaline than its iron counterpart from the Lewis definition of acid and base, and therefore cobalt oxide is more readily to form ternary compounds ($CoTiO_3$, Co_2TiO_4) with TiO_2. The solubility limit revealed by Raman spectroscopy is consistent with that observed via hyper-powder XRD analysis, which indeed shows the already formation of $CoTiO_3$ at 3 at% of Co^{2+} addition.

The oxidation states of cobalt and titanium in the TiO_2:Co powder are analyzed via high resolution XPS (Figure 44), using the 2 at% Co-doped sample as an example. The binding energies are referenced to the O-1s core level of TiO_2 (530.5 eV) to compensate for any possible electrostatic shifts caused by charging of the samples during photoelectron measurements. The Ti2p peaks are found at 458.1 and 463.8 eV (Figure 44a) with a spin-orbital doublet splitting (Δ=Ti2$p_{3/2}$-Ti2$p_{1/2}$) of 5.7 eV, suggesting an oxidation state of 4+ for titanium. The metallic Co2$p_{3/2}$ peak, reportedly occurs at 778 eV with Δ=14.86 eV, is not observed in the high-resolution XPS spectrum (Figure 44b). This is consistent with the results of hyper power XRD analysis (Figure 41), which neither shows the existence of Co metal phase up to a high doping level of 50 at%. The presence of strong shake-up satellites at 786.1 eV and

802.3 eV in the Co^{2+}-2p core level spectrum (Figure 44b) and the measured splitting of the $Co2p_{1/2}$-$Co2p_{3/2}$ orbital components Δ of 15.8 eV indicate that the Co atom in TiO_2 has an oxidation state of 2+ [33]. Besides, the occurrence of $Co^{II}Ti^{IV}O_3$ and $Co^{II}_2Ti^{IV}O_4$ compounds at higher doping levels provide additional evidence that cobalt has a valence of 2+ in the plasma synthesized nanopowders.

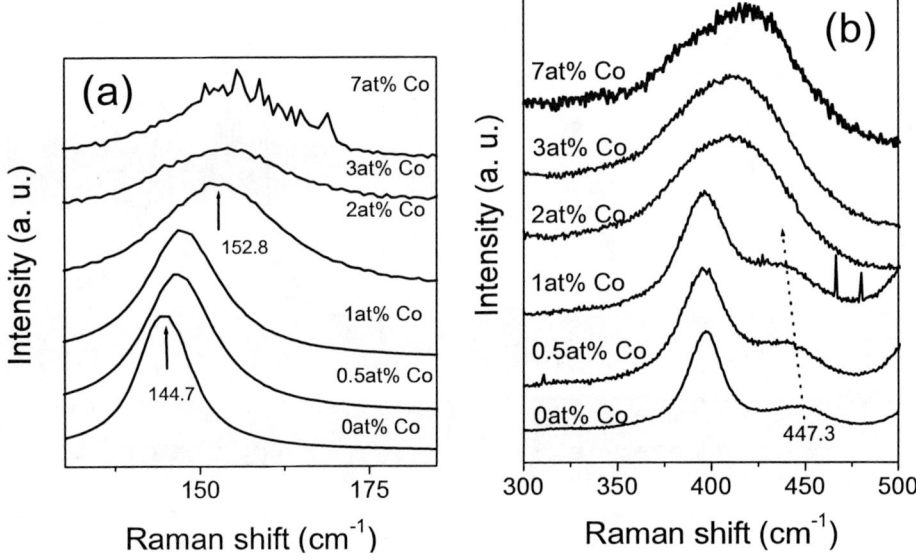

Figure 43. Raman spectra of the plasma synthesized nanopowders. Note the shifting of the anatase 144 cm^{-1} band (a) and the rutile 447 cm^{-1} band (b).

TEM observation (Figure 45a) indicates that the plasma generated nanoparticles (collected from the filter of the plasma reactor) are well dispersed, with sizes ranging from several nanometers to ~40 nm. HR-TEM analysis of lattice spacing indicates the presence of separated anatase (Figure 45b) and rutile (Figure 45c) nanocrystallites. Careful and massive TEM observations do not reveal the existence of any metallic Co clusters, in agreement with the results of XRD and XPS. The $CoTiO_3$ impurity presents itself as isolated nanocrystallites with well-developed twin structures (Figure 45d), making it readily distinguishable from the TiO_2 nanocrystallites. Extensive TEM observations indicate that the $CaTiO_3$ impurity nanocrystallites are hardly formed in the samples doped up to 2 at% of Co^{2+}

but appear with gradually increased frequencies at higher cobalt concentrations, further confirming that the solubility of Co^{2+} in the TiO_2 lattice is ~2 at%.

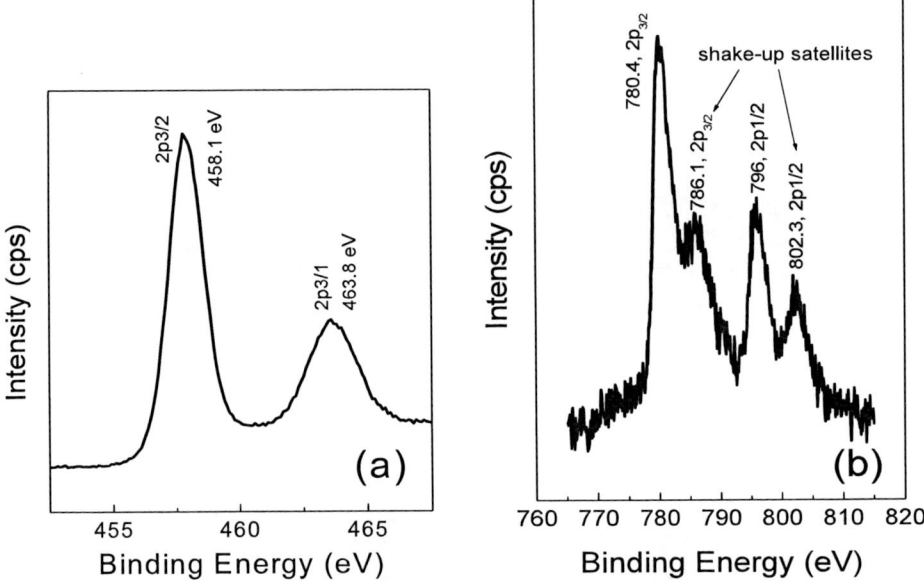

Figure 44. High resolution XPS spectra for Ti2p (a) and Co2p (b) of the 2 at% Co doped TiO_2 nanopowder.

Magnetic properties of the plasma-generated nanopowders are measured at 300 K, and the results are presented in Figure 46a. Normalization is made to reveal the contribution from each cobalt atom and the results are shown in Figure 46b. The existence of hysteresis loops clearly indicates that all the TiO_2:Co nanopowders made with the oxidative thermal plasma technique are ferromagnetic at room temperature. As no metallic Co or other ferromagnetic phase is detected by the various characterization techniques employed, it is thus plausible to conclude that the ferromagnetism is intrinsic of the samples. Nonetheless, the magnetization curves are apparently superposed contributions from the ferromagnetic and paramagnetic components, while the latter becomes predominant when the applied magnetic field is above about 3000 Oe. The ferromagnetic and paramagnetic components of magnetization

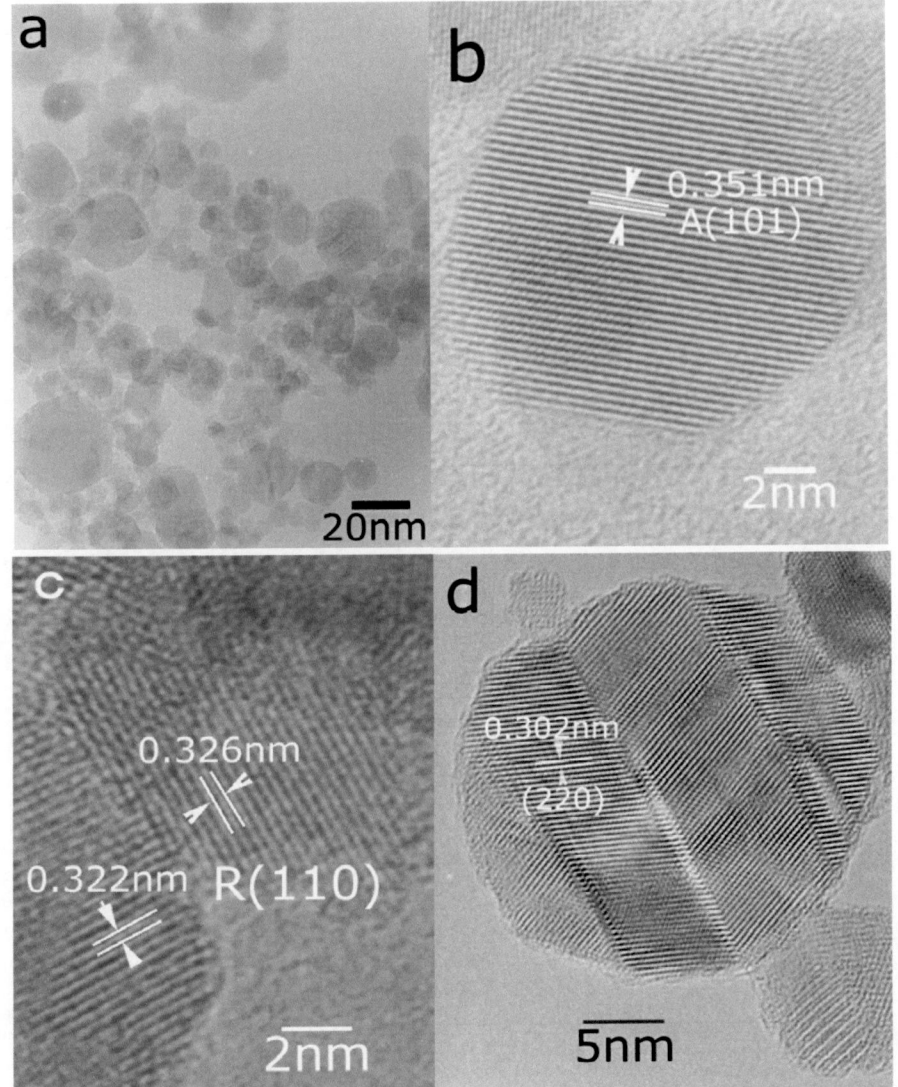

Figure 45. TEM analysis of the Co-doped TiO_2 nanoparticles, with (a): overall particle morphology of the 2 at% Co doped TiO_2; (b) and (c): lattice fringes of isolated anatase and rutile nanocrystallites observed in (a), respectively; (d): lattice image of a single $CoTiO_3$ nanocrystallite found in 10 at% Co-doped TiO_2.

can be separated from each other and the ferromagnetic parts are demonstrated in Figure 47, from which the saturation magnetization (Ms, Figure 48a), remnant magnetization (Mr, Figure 48b), and coercive force (Hc, Figure 48b) have been derived. It is observed that both the Ms and Mr tend to decrease along with increasing cobalt content in the samples, exhibiting deteriorated magnetic performances, while Hc shows weak dependence on the Co concentration (Figure 48). For DMS materials, one proposed mechanism for ferromagnetism is "carrier-mediated exchange", emphasizing the importance of carrier concentration and carrier mobility in the overall magnetic performances [37]. Doping TiO_2 with Co^{2+} steadily promotes the direct crystallization of rutile (Figure 41), in which the carriers are known to have much lower (1-2 order of magnitude) mobility than those in anatase. This may explain the decreased ferromagnetic performances at an increased Co^{2+} content. In addition, the formation of paramagnetic $CoTiO_3$ deteriorates further the normalized ferromagnetic properties of the samples.

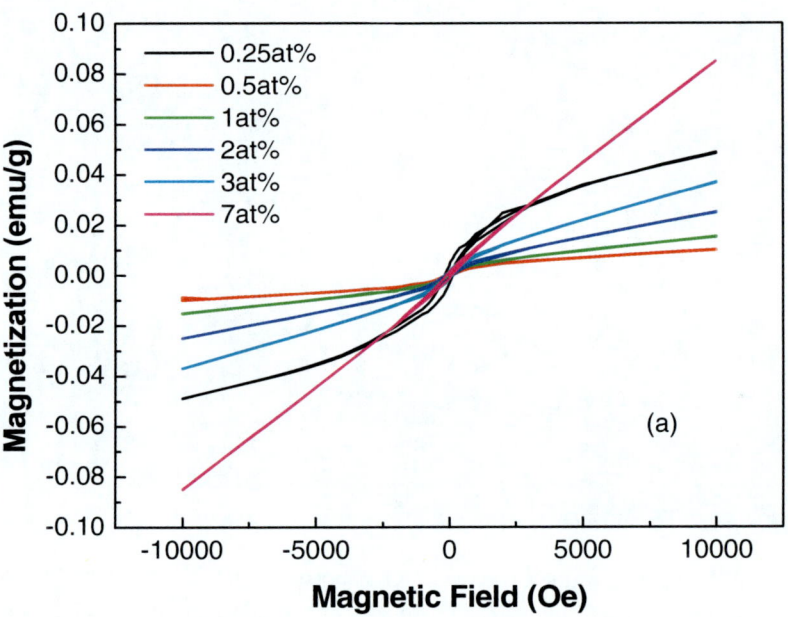

Figure 46. Continued on next page.

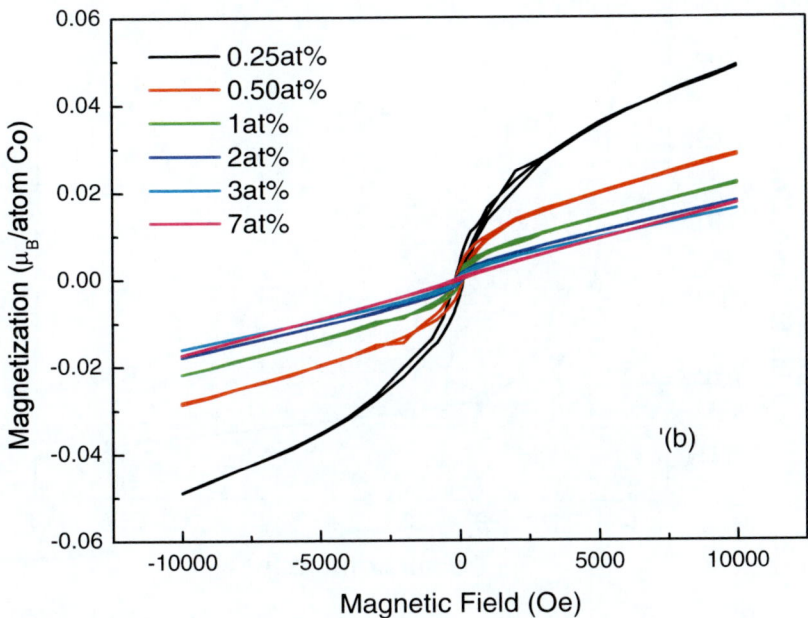

Figure 46. Magnetization curves of the Co-doped TiO$_2$ nanopowders measured at 300 K, with (a): the raw data, and (b): the normalized data.

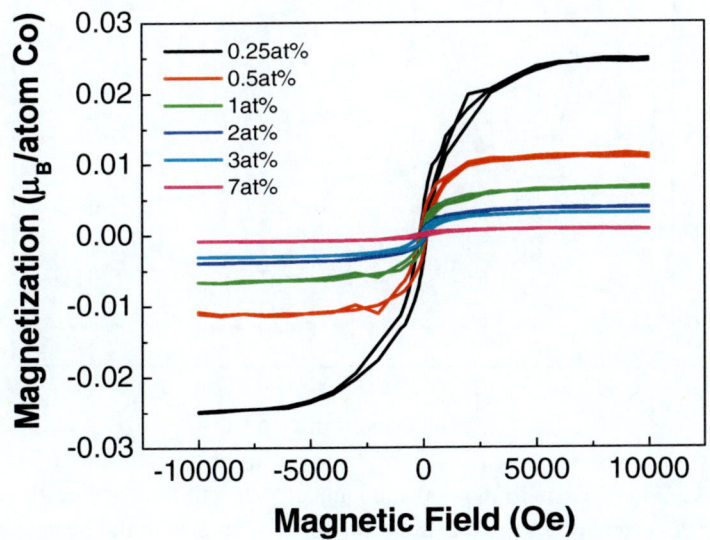

Figure 47. Ferromagnetic magnetization, derived from Figure 46, of the Co-doped TiO$_2$ nanopowders.

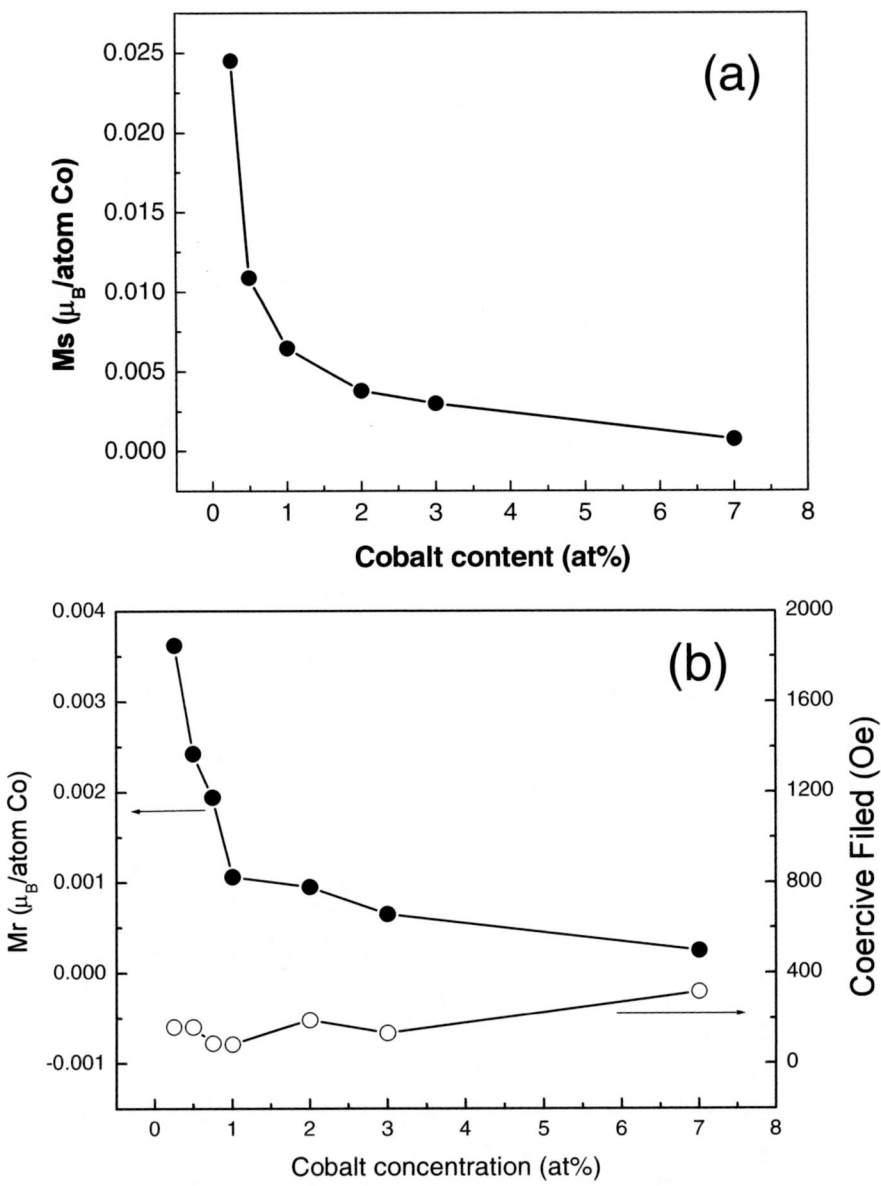

Figure 48. Saturation magnetization (a) and remnant magnetization and coercive force (b) of the TiO$_2$:Co nanopowders at room temperature, as a function of the Co content.

Chapter 7

CONCLUSION

Phase structure and morphology controlled synthesis of TiO_2 polymorphs (anatase, brookite, and rutile) as well as doping the TiO_2 nanocrystals with both non-metallic and metallic elements for novel/enhanced functionalities are demonstrated. As for solution processing via hydrothermal or via chemical precipitation under near-ambient conditions, it is shown that the phase selection of TiO_2 is largely determined by the reaction kinetics and is significantly affected by several processing parameters mainly including solution pH, reactant concentration, and the type of supporting anions present. It is argued that a fast reaction under high pH tends to yield anatase, a slow reaction under low pH favors rutile, while mild reaction conditions facilitate brookite crystallization. With this as a general guide, phase pure brookite, a polymorph of TiO_2 most difficult to obtain, can even be synthesized under near atmospheric conditions without the aid of any mineralizer. It is shown that the brookite clusters (~154 nm) transform to rutile at temperatures above 500 °C and that the phase transition, occurring within the individual clusters, does not involve the anatase polymorph and the transition kinetics is best described by a "contracting spherical interface" model. "Single molecular route" is proposed to be an efficient way to dope TiO_2 with nonmetallic elements for enhanced photocatalytic performances under visible light irradiation. With this strategy, the (C, N, Cl) codoped anatase nanocrystals pyrolyzed from the $Ti^{III}[OC(NH_2)_2]_6Cl_3$ coordination compound show substantially better photocatalytic reactivity than the well known Degussa P25 TiO_2 in the bleaching methyl orange solutions, either at a fixed TiO_2 loading or at a fixed surface area of the loaded TiO_2. Utilizing the extremely high processing temperature of the RF thermal plasma and the fast quenching at the tail part of the plasma flume, doping TiO_2 with both non-metallic and metallic elements for novel/enhanced functionality is achieved via one single step of processing. One issue that remains in thermal plasma processing is that phase selection of the anatase and rutile polymorphs has not yet been fully achieved, though this can be manipulated to some extent via the injection of additional quench gases. This is primarily due to the fast quenching at the plasma tail and the significantly differed thermal histories of the individual TiO_2 clusters in the plasma field. Like nitrogen doping, chlorine doping of TiO_2 via plasma processing may also effectively shift the optical response of TiO_2 into the visible light regime, rendering the material reactive under visible light illumination. Doping TiO_2 with transition metal (Co^{2+}) for DMS application is presented, and it is concluded that the room temperature ferromagnetism is intrinsic of the nanocrystals synthesized via thermal plasma. Efficient non-radiative energy transfer from the TiO_2 host lattice to activator ions (Eu^{3+}, Er^{3+}) is confirmed in the nanocrystals processed via RF thermal plasma, and as a result excellent luminescence is achieved either by exciting the TiO_2 host (above the bandgap) or by directly exciting the activator ions (below the bandgap). Such a luminescent property would allow the rare-earth doped TiO_2 nanocrystals to find applications in optoelectronic devices.

ACKNOWLEDGMENT

The studies have been partially supported by the Program for New Century Excellent Talents in University (Grant NCET-25-0290), the National Science Fund for Distinguished Young Scholars (Grant 50425413), the Program for Changjiang Scholars and Innovative Research Teams in University (PCSIRT, IRT0713), and the National Natural Science Foundation of China (Grants 50672014 and 50772020). Thanks are also due to our colleagues and students who have contributed to the technical contents.

REFERENCES

[1] Tompsett, G. A.; Bowmaker, G. A.; Cooney, R. P.; Metson, J. B.;Rogers, K. A.; Seakins, J. M. *J. Raman Spectrosc.* 1995, *26*, 57-62.
[2] Chaves, A.; Katiyar. R. S.; Porto, S. P. S. *Phys. Rev B* 1974, *10*, 3522-3533.
[3] Ohsaka, T.; Izumi, F.; Fujiki, Y. *J. Raman Spectrosc.* 1978, *7*, 321-324.
[4] Parker, J. C.; Siegel, R. W. *Appl. Phys. Lett.* 1990, *57*, 943-945.
[5] Chen, X.; Mao, S. S. *Chem. Rev.* 2007, *107*, 2891-2959.
[6] Cheng, H.; Ma, J.; Zhao, Z.; Qi, L. *Chem. Mater.* 1995, *7*, 663-671.
[7] Li, J.-G.; Tang, C.; Li, D.; Haneda, H.; Ishigaki, T. *J. Am. Ceram. Soc.* 2004, *87*, 1358-1361.
[8] Li. J.-G.; Sun, X. D.; Ishigaki, T. *J. Phys. Chem. C* 2007, *111*, 4969-4976.
[9] Kominami, H.; Kohno, M.; Kera, Y. *J. Mater. Chem.* 2000, *10*, 1151-1156.
[10] Henry, M.; Jolivet, J. P.; Livage, J. in *Aqueous Chemistry of Metal Cations, Hydrolysis, Condensation, and Complexation*; Reisfeld, R.; Jorgensen C. K. Ed.; Springer-Verlag, Berlin, 1992, p. 155.
[11] Pottier, A.; Chaneac, C.; Tronc, E.; Mazerlolles,L.; Jolivet, J. P. *J. Mater. Chem.* 2001, *11*, 1116-1121.
[12] Zhang, H. Z.; Banfield, J. F. *J. Phys. Chem. B* 2000, *104*, 3481-3487.

[13] Hoffmann, M. R.; Martin, S. T.; Choi, W. Y.; Bahnemann, D. W. *Chem. Rev.* 1995, *95*, 69-96.
[14] Wu, M.; Lin, G.; Chen, D.; Wang, G.; He, D.; Feng, S.; Xu, R. *Chem. Mater.* 2002, *14*, 1974-1980.
[15] Aruna, S. T.; Tiroshi, S.; Zaban, A. *J. Mater. Chem.* 2000, *10*, 2388-2391.
[16] Li, Y. L.; Ishigaki, T. *J. Phys. Chem. B* 2004, *108*, 15536-15542.
[17] Morgan, D. L.; Zhu, H. Y.; Frost, R. L.; Waclawik, E. R. *Chem. Mater.* 2008, *20*, 3800-3802.
[18] Privman, V.; Goia, D. V.; Park, J.; Matijevic, E. *J. Colloid Interface Sci.,* 1999, *213*, 36-45.
[19] Li, J.-G.; Ishigaki, T. *Acta Mater.* 2004, *52*, 5143-5150.
[20] Liu, S. H.; Sun, X. D.; Li, J.-G.; Li, X. D.; Xiu, Z. M.; Ho, D. *Eur. J. Inorg. Chem.* 2009, 1214-1218.
[21] Asahi, R.; Morikawa, T.; Ohwaki, T.; Aoki, K.; Taga, Y. *Science* 2001, *293*, 269-271.
[22] Li, J.-G.; Yang, X. J.; Ishigaki, T. *J. Phys. Chem. B* 2006, *110*, 14611-14618.
[23] Boulos, M. I.; Fauchais, P.; Phender, E. *Thermal Plasmas: Fundamental and Applications*, , Plenum Press: NY, 1994; Vol. 1, p. 33.
[24] Li, J.-G.; Kamiyama, H.; Wang, X. H.; Moriyoshi, Y.; Ishigaki, T. *J. Eur. Ceram. Soc.* 2006, *26*, 423-428.
[25] Oh, S.-M.; Li, J.-G.; Ishigaki, T. *J. Mater. Res.* 2005, *10*, 529-537.
[26] Li, J.-G.; Ikeda, M.; Ye, R.; Moriyoshi, Y.; Ishigaki, T. *Appl. Phys. D: Appl. Phys.* 2007, *40*, 2348-2353.
[27] Li, J.-G.; Ikeda, M.; Tang, C.; Moriyoshi, Y.; Hamanaka, H.; Ishigaki, T. *J. Phys. Chem. C* 2007, *111*, 18018-18024.
[28] Li, J.-G.; Wang, X. H.; Watanabe, K.; Ishigaki, T. *J. Phys. Chem. B* 2006, *110*, 1121-1127.
[29] Wang, X. H.; Li, J.-G.; Kamiyama, H.; Katada, M.; Ohashi, N.; Moriyoshi, Y.; Ishigaki, T. *J. Am. Chem. Soc.* 2005, *127*, 10982-10990.
[30] Li, J.-G.; Wang, X. H.; Tang, C.; Ishigaki, T.; Tanaka, S. *J. Am. Ceram. Soc.* 2008, *91*, 2032-2035.

[31] Frindell, K. L.; Bartl, M. H.; Robinson, M. R.; Bazan G. C.; Popitsch, A.; Stucky, G. D. *J. Solid State Chem.* 2003, *172*, 81-88.
[32] Matsumoto ,Y.; Murakami, M.; Shono, T.; Hasegawa, T.; Fukumura, T.; Kawasaki, M.; Ahmet, P.; Chikyow, T.; Koshihara, S.; Koinuma, H. *Science* 2001, *291*, 854-856.
[33] Li, J.-G.; Buchel, R.; Isobe, M.; Mori, T.; Ishigaki, T. *J. Phys. Chem. C* 2009, *113*, 8009-8015.
[34] Matsumoto, Y.; Takahashi, R.; Murakami, M.; Koida, T.; Fan, X. J.; Hasegawa, T.; Fukumura, T.; Kawasaki, M.; Koshihara, S. Y.; Koinuma, H. *Jpn. J. Appl. Phys.* 2001, *40*, L1204.
[35] Park, W. K.; Ortega-Hertogs, R. J.; Moodera, J. S.; Punnoose, A.;Seehra, M. S. *J. Appl. Phys.* 2002, *91*, 8093-8095.
[36] Joh, Y. G.; Kim, H. D.; Kim, B. Y.; Woo, S. I.; Moon, S. H.; Cho, J. H.; Kim, E. C.; Kim, D. H.; Cho, C. R. *J. Korean Phys. Soc.* 2004, *44*, 360-367.
[37] Fukumura, T.; Toyosaki, H.; Yamada, Y. *Semicond. Sci. Technol.* 2005, *20*, S103-S111.

INDEX

A

absorption, 11, 12, 13, 40, 45, 46, 47, 57, 58, 59, 60, 64, 65
absorption coefficient, 11, 12
absorption spectra, 11, 12, 45, 46, 58, 59, 64
acid, 6, 15, 16, 20, 22, 61, 74
acidity, 18, 20, 22
activation, 29, 31, 32
activation energy, 31, 32
activators, 65, 69, 70
adjustment, 6, 25
adsorption, 12
aggregation, 26, 34, 37
aid, 27, 81
air, 41, 42, 43, 46
alkaline, 16, 35, 74
ambient pressure, 26
ammonia, 36, 44, 45, 62, 68
ammonium, 6, 7, 33, 34, 35, 36
ammonium hydroxide, 35, 36
amorphous, 10, 15, 18, 33
anatase, xi, 1, 2, 3, 5, 7, 8, 9, 10, 11, 12, 13, 16, 17, 18, 20, 22, 23, 25, 26, 28, 29, 32, 33, 34, 36, 42, 43, 45, 47, 52, 53, 56, 57, 58, 62, 63, 64, 68, 71, 72, 74, 75, 77, 78, 81
annealing, 16, 26, 27, 28, 29, 31, 32, 40, 43, 46, 52, 69
application, 3, 29, 39, 82
aqueous solution, 6, 9, 12, 13, 22, 25, 61
aqueous solutions, 6, 13
aqueous suspension, 16
Arrhenius law, 31
aspect ratio, 21
assumptions, 29
asymmetry, 65
atmosphere, 29, 40, 59, 71
atmospheric pressure, 33

B

bandgap, 12, 14, 39, 57, 64, 70, 82
benchmark, 47
binding, 74
binding energies, 74
bleaching, 13, 41, 46, 47, 59, 82
blocks, 22
building blocks, 22

C

calibration, 53
carbon, 36, 44
carbon dioxide, 36
carboxylic acids, 5
carrier, 40, 78
cation, 9, 22
chemical composition, 43, 49, 53
chemicals, 61
chlorine, 10, 53, 54, 57, 82
clustering, 26
clusters, 26, 30, 31, 63, 71, 75, 81
Co, 50, 71, 72, 74, 75, 76, 77, 79, 80
CO2, 15, 44
cobalt, 70, 71, 73, 74, 76
combined effect, 20
compensation, 63, 72, 74
components, 75, 76
composites, 57
composition, 53, 66
compounds, 5, 40, 43, 71, 74, 75
concentration, 5, 9, 10, 12, 13, 27, 29, 40, 45, 47, 69, 78, 81
condensation, 9
constant rate, 18
control, 25, 49, 50
conversion, 16, 17, 18, 19, 22, 43
cooling, 49, 53
correlation, 3
cosmetics, 1, 3, 24
crystal growth, 15
crystal structure, 1, 2, 3, 6, 9, 15, 22, 43, 57, 72
crystal structures, 1, 2, 57
crystalline, 1, 3, 57
crystallinity, 16, 43, 57, 69
crystallites, 3, 6, 7, 10, 12, 16, 19, 22, 23, 26, 29, 34, 43, 53, 57, 63, 64
crystallization, xi, 5, 9, 10, 12, 15, 18, 23, 25, 27, 29, 34, 63, 72, 78, 81

crystals, 10, 22, 26, 35

D

decomposition, 43
defects, 70
deficiency, 3, 15, 24, 53, 58
degradation, 13, 40
Degussa, xi, 16, 41, 44, 45, 57, 82
density, 2, 3, 26, 36
deposition, 72
desorption, 43, 45, 54
destruction, 14, 47
diffraction, 29, 63, 64
diffusion, 26
dispersion, 15, 57
distilled water, 62, 68
distribution, 29, 40, 55, 63
dopant, 13, 40, 41, 44, 45, 47, 62, 71
dopants, 43, 57, 69
doped, 42, 50, 61, 63, 65, 66, 67, 68, 69, 70, 74, 75, 76, 77, 79, 82
doping, xi, 3, 39, 41, 44, 49, 55, 70, 71, 72, 74, 81

E

earth, 70, 82
electromagnetic, 45
electron, 47
emission, 65, 66, 68, 69
energy, 6, 13, 14, 39, 65, 67, 69, 70, 82
energy transfer, 65, 67, 69, 70, 82
enlargement, 64
epitaxial growth, 19, 20
epitaxy, 72
erbium, 68, 69
ethanol, 55
europium, 61, 62
evaporation, 52
excitation, 65, 66, 67, 69, 70

expert, viii

F

ferromagnetism, 70, 76, 82
film, 71, 72
films, 36, 70
flame, 16, 49, 53, 55
flow, 51, 52, 53, 57, 58, 59, 62, 63
flow rate, 51, 52, 53, 57, 58, 59
free energy, 53
FTIR, 40, 43
FWHM, 69

G

gas, 15, 40, 49, 51, 53, 57, 63, 68
gases, 49, 82
generation, 6, 51
grains, 15
growth, 7, 9, 10, 12, 15, 19, 20, 22, 26, 27, 35, 36, 53
growth rate, 36

H

H_2, 15, 72
harvesting, 39
heating, 31, 45
heating rate, 45
high resolution, 74
high temperature, xi
host, 64, 65, 67, 69, 70, 82
hydrogen, 6, 7
hydrogen peroxide, 6, 7
hydrolysis, 9, 26, 34, 36, 61
hydrolyzed, 9
hydrothermal, ix, xi, 5, 10, 12, 15, 16, 17, 18, 19, 20, 21, 22, 23, 25, 26, 40, 81
hydrothermal process, 25

hydrothermal synthesis, 15
hydroxide, 35, 36
hysteresis, 76
hysteresis loop, 76

I

identification, 3
illumination, 40, 59, 82
images, 7, 20, 21, 35
induction, 50
infinite, 30
injection, 82
interactions, 9, 10, 13
interface, 30, 32, 81
ions, xi, 6, 10, 22, 26, 27, 43, 59, 60, 63, 64, 65, 69, 70, 72, 74, 82
iron, 74
irradiation, 14, 39, 41, 46, 47, 60, 61, 81

K

kinetic model, 26, 29
kinetics, 13, 14, 25, 29, 61, 81
Kolmogorov, 29
Korean, 87

L

lanthanide, xi
laser, 65, 67
lattice, 3, 6, 7, 39, 42, 43, 44, 45, 53, 57, 58, 63, 64, 65, 67, 69, 70, 71, 74, 75, 77, 82
lattices, 53
lifetime, 3
ligands, 12
linear, 1, 9, 13, 18, 22, 32
linear regression, 18
luminescence, 65, 69, 82

M

magnetic field, 76
magnetic properties, 71
magnetization, 76, 79, 80
magnetoelectronics, 70
mass transfer, 12
MBE, 72
melting, 53
microwave, 70
migration, 13
mixing, 41, 47, 62
mobility, 78
models, 2, 29
moisture, 6
molar ratio, 6, 17, 18, 20, 21, 35, 36, 61, 62, 72
molar ratios, 62
mole, 54
molecular beam epitaxy, 72
molecules, 43
moon, 87
morphology, xi, 3, 6, 14, 26, 29, 37, 55, 63, 77, 81
movement, 30

N

NaCl, 5
nanocrystalline, 43
nanocrystals, xi, 3, 5, 8, 14, 15, 16, 23, 25, 33, 39, 41, 42, 49, 50, 70, 74, 81
nanomaterials, xi, 1, 3, 37
nanometers, 75
nanoparticles, 26, 34, 35, 36, 49, 53, 65, 70, 75, 77
nanorods, 6, 13, 15, 20, 23
nanostructures, 16
Ni, 71
nitrate, 61, 68, 71
nitric acid, 6, 15, 16, 22

nitrogen, 39, 43, 44, 82
non-linearity, 30
non-uniform, 40
nucleation, 10, 20, 26, 27, 29, 35, 36, 53
nuclei, 20, 37

O

observations, 18, 55, 71, 75
optical, 2, 12, 39, 82
optoelectronic devices, 70, 82
optoelectronics, 70
oxidation, xi, 6, 15, 26, 27, 35, 40, 51, 52, 71, 72, 74
oxidation products, 52
oxidative, 71, 76
oxide, 71, 74
oxides, 70, 71
oxygen, 3, 6, 8, 15, 24, 26, 27, 43, 53, 57, 58, 59, 63, 70, 72, 74

P

parameter, 31
particle morphology, 29, 77
particles, 15, 20, 26, 28, 29, 30, 32, 34, 35, 36, 37, 49, 55, 56, 57, 58, 59, 63
pathways, 57
pH, 5, 7, 9, 10, 12, 15, 16, 18, 22, 25, 27, 33, 36, 62, 68, 81
phase diagram, 16
phase transformation, xi, 29
phonon, 3
photocatalysis, 1, 39, 47
photocatalysts, 14, 39, 61
photoluminescence, 70
photoluminescent, 61
photon, 12
photovoltaic cells, 1, 39
physical properties, 1, 70
physicochemical properties, xi

Index

pigments, 1, 3, 24, 39
plasma, xi, 15, 49, 50, 51, 52, 53, 54, 55, 57, 58, 59, 60, 61, 62, 63, 68, 69, 70, 71, 72, 73, 75, 76, 82
PLD, 72
powder, 13, 16, 19, 20, 22, 26, 27, 29, 42, 44, 45, 47, 50, 51, 53, 54, 55, 56, 57, 58, 60, 63, 64, 68, 69, 71, 72, 74
powders, 7, 8, 11, 12, 13, 21, 23, 29, 30, 31, 40, 43, 44, 45, 46, 47, 49, 51, 52, 53, 54, 56, 57, 58, 59, 62, 63, 69, 72, 74
power, 50, 51, 73, 74
precipitation, 36, 70, 81
preference, 35
pressure, 26, 33, 50, 51, 59
pristine, 39
probability, 9, 20
probe, 51
property, viii, 3, 82
pulsed laser, 72
pulsed laser deposition, 72
purification, 49
pyrolysis, 16, 49, 53, 55, 72

R

Raman, 2, 8, 12, 23, 26, 52, 53, 57, 58, 74, 75, 85
Raman scattering, 2
Raman spectra, 8, 23, 52, 53, 74, 75
Raman spectroscopy, 3, 8, 12, 26, 57, 58, 74
random, 29
range, 5, 10, 29, 36, 39, 44, 45, 49, 51, 53, 54, 58, 60, 63, 65, 69
reactant, 5, 37, 81
reaction rate, 25
reaction temperature, 9, 34
reaction time, 5
reactivity, 3, 60, 82
reciprocal temperature, 33
recombination, 40, 47, 60

recrystallization, 10, 16
redox, 12, 26
refractive index, 1, 3, 24
refractory, 49
regression, 18
relationship, 31, 32
resin, 69
resolution, 74, 76
rice, 35, 36
rods, 20
room temperature, 68, 70, 71, 76, 80, 82
RTF, 70, 72
rutile, xi, 1, 2, 3, 5, 7, 8, 9, 11, 12, 13, 15, 16, 17, 18, 19, 20, 22, 23, 25, 26, 27, 28, 29, 30, 31, 32, 36, 45, 47, 52, 53, 56, 57, 58, 62, 63, 64, 68, 71, 72, 74, 75, 77, 78, 81

S

sample, 13, 34, 45, 46, 47, 56, 59, 64, 65, 66, 68, 69, 71, 74
saturation, 19, 77
scattering, 3, 23
sedimentation, 55
selected area electron diffraction, 6
SEM, 26, 34, 35, 42, 43
SEM micrographs, 34
semiconductor, 12
semiconductors, 70
separation, 13
services, viii
shape, 3, 13, 26, 36
signals, 54, 67
silicon, 69
sintering, 26
SiO_2, 69
sites, 39, 72, 74
sodium, 35, 36
sodium hydroxide, 35, 36
solar, 39
solar energy, 39

sol-gel, 5, 32, 40, 72
solid state, 72
solubility, 20, 22, 63, 68, 71, 74, 76
species, 9, 20, 22, 43, 45, 55
specific surface, 6, 13, 16, 21, 26, 47, 59, 72
spectroscopy, 13, 35, 43, 57
spectrum, 3, 43, 45, 53, 65, 66, 74
speculation, 71
spheres, 19, 26, 29, 33, 34, 37, 57
sputtering, 40, 72
stability, 6, 10, 39
stabilization, 20, 27
stabilize, 9
stainless steel, 50
steel, 50
stoichiometry, 3, 53, 57
strategies, 40
structure formation, 10
students, 83
substitutes, 74
substrates, 36
sun, vii, 85, 86
supply, 50
surface area, 7, 13, 41, 46, 47, 53, 59, 82
surface properties, 59
surface water, 44, 54
suspensions, 36
symmetry, 9
synthesis, xi, 3, 5, 15, 50, 53, 60, 81

T

TEM, 6, 7, 20, 21, 26, 42, 43, 55, 57, 58, 63, 75, 77
temperature, xi, 16, 19, 27, 29, 31, 33, 40, 44, 45, 46, 49, 52, 54, 68, 71, 82
thermal decomposition, 43
thermal plasma, xi, 15, 49, 50, 51, 52, 55, 60, 61, 68, 71, 76, 82
thermal stability, 39
thin film, 70

titanates, 72
titania, xi, 16
titanium, 1, 5, 6, 9, 10, 12, 15, 20, 22, 25, 33, 35, 40, 41, 43, 61, 68, 71, 74
titanium alkoxide, 34
titanium isopropoxide, 15
titration, 35, 36
tolerance, 74
toxicity, 39
tracking, 66
trajectory, 53
transfer, 65, 67, 69, 70, 82
transformation, xi, 17, 19, 20, 29, 32, 45
transition, xi, 11, 13, 16, 26, 29, 31, 32, 39, 65, 67, 70, 81
transition metal, 70, 82
transitions, 13, 64, 65

U

uniform, 29, 36, 40, 41, 43
urea, 6, 25, 27, 33, 34, 36, 40, 41, 43
UV, 11, 12, 13, 14, 35, 39, 45, 46, 57, 59, 61, 64, 65, 70
UV absorption, 60
UV irradiation, 14, 60
UV light, 13, 39, 59, 70

V

vacancies, 3, 53, 63, 70, 72
vacuum, 44, 50, 71, 72
valence, 75
values, 12, 16, 29, 57
Van der Waals, 26
vibration, 41, 43

W

water, 36, 40, 44, 50, 61, 68

wavelengths, 58
weakness, 64

X

xerogels, 69
XPS, 74, 75, 76
X-ray analysis, 36
XRD, 17, 18, 27, 29, 30, 36, 43, 51, 52, 53, 56, 62, 69, 71, 73, 74, 75

Y

yield, 6, 9, 34, 53, 62, 81

Z

ZnO, 71